Protein purification
applications

Unless

The Practical Approach Series

Related **Practical Approach** Series Titles

Protein purification applications

Second Edition

A Practical Approach

Edited by

Simon Roe

AEA Technology plc, Bioprocessing Facility,
Transport Way, Watlington Road,
Oxford OX4 6LY, U.K.

OXFORD
UNIVERSITY PRESS

OXFORD

UNIVERSITY PRESS

Great Clarendon Street, Oxford OX2 6DP

Oxford University Press is a department of the University of Oxford.
It furthers the University's objective of excellence in research,
scholarship, and education by publishing worldwide in

Oxford New York

Athens Auckland Bangkok Bogotá Buenos Aires Calcutta Cape Town
Chennai Dar es Salaam Delhi Florence Hong Kong Istanbul Karachi
Kuala Lumpur Madrid Melbourne Mexico City Mumbai Nairobi Paris
São Paulo Singapore Taipei Tokyo Toronto Warsaw

with associated companies in Berlin Ibadan

Oxford is a registered trade mark of Oxford University Press in the UK
and in certain other countries

Published in the United States by Oxford University Press Inc., New York

First edition published 1990
Reprinted 1990, 1993, 1995
Second edition published 2001

A catalogue record for this title is available from the British Library

Library of Congress Cataloging in Publication Data
Protein purification applications a practical approach / edited by
Simon Roe. – 2nd ed.
(Practical approach series ; 245)
Includes bibliographical references and index.
1. Proteins–Purification–Laboratory manuals.   I. Roe, Simon.
II. Series
QP551.P6975135 2001   572.6–dc21   00-048306

1 3 5 7 9 10 8 6 4 2

ISBN 0 19 963672 9 (Hbk.)
ISBN 0 19 963671 0 (Pbk.)

Typeset in Swift by Footnote Graphics, Warminster, Wilts
Printed in Great Britain on acid-free paper
by The Bath Press, Avon

Preface

Biotechnology is now playing an increasingly important role in our lives. The last 20 years has seen an explosion in the application of genetics, cells, and their components to improve the quality of life, from new therapeutic treatments and rapid diagnostic tests to improved disease resistance in crops. At the time of writing, some 1300 biotechnology companies have emerged in the USA employing over 150 000 people, with new therapeutic products providing multi-million dollar sales. This same trend is now being repeated in Europe and the next five years should see the successful commercialization of new biotechnology companies and the transition of biotechnology start-ups into established players.

Such commercial progress is, of course, the result of our improved understanding of the life sciences and the causes of human disease. While new developments in cell biology, immunology, and genetics have been widely publicized, the significant advances in separations technology have played a crucial role in improving our knowledge of the life sciences. Since the early days of biochemistry, when the fundamentals of cell metabolism and enzyme function were being established, protein separation techniques have been an essential part of the biotechnology revolution. Any biochemistry laboratory, whether in schools, universities, high-tech biotechnology companies, or pharmaceuticals giants, will invariably contain the essential tools of the trade—centrifuges, spectrophotometers, buffers, and chromatography columns. In fact proteins are perhaps the most widely purified type of biological molecule since they form an integral part of cellular metabolism, structure, and function. Protein purification has continued to evolve during the last ten years, with improvements in equipment control, automation, separations materials, and the introduction of new protein separation techniques such as affinity membranes and expanded beds. Microseparation techniques have emerged for protein sequencing and the highly sophisticated (and often expensive) equipment now available allows chromatography process development in hours where days or even weeks might have been required in the past.

These advances in protein separation are likely to continue, driven by the need for faster process development and improved resolution. Such developments have clearly reduced the workload but may have removed the purification scientist from an understanding of the fundamentals of the science of purification. Although sophisticated equipment may now ease the workload, there is still a need to consider how purification strategies are designed and unit operations joined together to produce a processing train which can, if required, be scaled-up efficiently. In addition, an understanding of protein purification facilitates problem solving and enables the design of a robust process for production purposes.

As a result, it is appropriate that two new editions of '*Practical approaches in protein purification*' should appear, covering protein purification techniques, and protein purification applications. The focus of these publications continues to be on the provision of detailed practical guidelines for purifying proteins, particularly at the laboratory scale and, where possible, information is summarized in tables and recipes. The books are primarily aimed at the laboratory worker and assume an understanding of the basics of biochemistry. Such are the advances in protein purification and the widespread use of the techniques that the two volumes cannot be comprehensive. Nevertheless each of the key techniques used at a laboratory scale is covered, starting from an overview of purification strategy and analytical techniques, and followed by initial extraction and clarification techniques. Since chromatography forms the backbone of modern purifications strategies, several chapters are devoted to this technique. As with the previous editions, several chapters also cover specific considerations involved in the purification from certain protein sources. While the emphasis of the books are on laboratory scale operations, where relevant, information is included within the appropriate chapter on scale-up considerations, and one is devoted to this area in more detail. Given the wide variety of knowledge involved, each chapter has been written by an author who is a specialist on the given subject.

Although successful protein purification requires an understanding of both biochemistry and the science of separation, I have always thought that the process of designing a purification strategy has an artistic element involved and is a challenge to be enjoyed rather than laboured over. Starting from a crude mixture of proteins, carbohydrates, lipids, and cell debris, a well thought out sequence of steps leading ultimately to a highly pure single protein is a work of art which must be admired. I hope that the new editions kindle the intense interest in protein purification which I have enjoyed over the years. Who knows, even those who have long hung up their laboratory coats may head down to the labs in the late hours to potter around with the odd ion exchange column. My apologies to their partners in advance!

Didcot   S. D. R.
November 2000

Contents

Protocol list

Abbreviations

aa	amino acids
APS	ammonium persulfate
α_{S1}–CN	alpha S1 casein
α_{S2}–CN	alpha S2 casein
α-La	alpha lactalbumin
Bis	*N,N'*-methylene-bis-acrylamide
β-CN	beta casein
β-LG	beta lactoglobulin
BSA	bovine serum albumin
CBD	cellulose binding site
CDTA	cyclohexane diamino tetraacetic acid
CHAPS	3-[3-(cholamidopropryl) dimethylammonia]-1-propane sulfonate
ConA	concanavalin
CV	column volume
3,4 DC	3,4-dichlorisocoumarin
DEAE	diethylaminoethyl
Dip-F	di-isoprophyl flurophosphate
DTT	dithiothreitol
E-64	trans-epoxy succinyl-L-leucylamido-(4-guanidino) butane
EDTA	ethylenediaminetetraacetic acid
EST	expressed sequence tag
EWL	egg white lysozyme
FCS	fetal calf serum
GAPDH	glyceraldehyde 3-phosphate dehydrogenase
GPI	glycosyl-phosphatidylinositol
GSH	reduced glutathione
GST	glutathione transferase
HCG	human chorionic gonadotrophin
HIC	hydrophobic interaction chromatography
IA	immobilized antibody
IEC	ion exchange chromatography
IEF	isoelectric focusing
Ig	immunoglobulin

I-IgG	immobilized immunoglobulin G
IMAC	immobilized metal affinity chromatography
IS	immobilized substrate
κ-CN	kappa casein
LC-MS	liquid chromatography-mass spectrometry
MCS	multiple cloning site
MES	2(*N*-morpholine) ethane sulfonic acid
MS-MS	tandem mass spectrometry
PAGE	polyacrylamide gel electrophoresis
PAL	phenylalanine ammonia lyase
PCR	polymerase chain reaction
PEG	polyethylene glycol
PGK	3-phosphoglycerate kinase
PI-PLC	phosphatidylinositol-specific phospholipase
PMSF	phenylmethylsulfonyl fluoride
pptn	selective precipitation
PVP	polyvinylpyrrolidone
RME	reverse micellar extraction
RP	reverse-phase
RPAS	recombinant phage antibody system
RP-HPLC	reverse-phase high performance liquid chromatography
RUBISCO	ribulose bisphosphate carboxylase/oxygenase
SA	serum albumin
SBD	starch binding domain
SDS	sodium dodecyl sulfate
SMUF	simulated milk ultra filtrate
SPA	streptococcal protein A
SPG	streptococcal protein G
TEMED	*N,N,N′,N′*,-tetramethyethylenediamine
TFA	trifluoroacetic acid
TIM	triosephosphate isomerase

Chapter 1

Fusion protein purification methods

Stephen J. Brewer

Bioproducts Technology Consultant, 'Poldhune' Parc Owles, St. Ives, Cornwall TR26 2RE, UK.

Charles E. Glatz

Department of Chemical Engineering, 2114 Sweeney Hall, Iowa State University, Ames, IA 50011–2230, USA.

Charles Dickerson

AEA Technology plc, Didcot, Oxfordshire OX11 ORA, UK.

1 Introduction

Genetic engineering has provided scientists and technologists with the opportunity to change the properties of proteins so that their purification properties can be improved. By the insertion of DNA sequences on either the 3' or 5' ends of translated DNA, the amino acid sequences on either end of a protein can be changed to provide a purification fusion. These fusions may be used to simply stabilize the protein from attack by proteases by helping the protein form an inclusion body. The fusion might also impart specific purification properties on the protein, making it suitable for immunoaffinity, metal chelate, ion exchange, hydrophobic chromatography, or partition. Alternatively, for well-characterized proteins, specific amino acids may be changed within the protein to introduce patches with a specific affinity for an adsorption matrix.

Many of the practical aspects of this work have been reviewed previously in the Practical Approaches series (1). Since then, developments have occurred in the field of process-scale technology, particularly in the use of preparative HPLC and polymer precipitation methods. There is also a need to address the analytical methods required to follow the digestion of protein fusions and the methods needed to remove the product of digestion. In addition, a number of commercially available fusion kits are now available. These will be of particular application to those needing to rapidly purify the proteins which are emerging from gene-sequencing programs and genomics technology. Therefore, in this up-date on the previous chapter, we have attempted to emphasize these new advances as well as reviewing the general principles of purification fusion technology.

The principle of purification fusions is very straightforward (*Figure 1*). However, each protein is unique, and the design of the protein fusion will need inputs

1

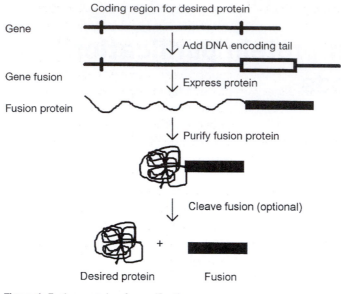

Figure 1 Fusion proteins for purification.

from genetic engineering and from a protein engineering standpoint. When the protein is being used as a bioanalytical reagent or as a biocatalyst, then as long as the fusion does not interfere with the activity of a protein, it is quite acceptable to leave it on the protein. If, however, the protein is to be used as a long-term therapeutic product, then the fusion protein will probably need to be removed in order to prevent an adverse immunological response. Therefore, in considering which purification fusion to use, the ultimate end use of the protein needs to be considered.

This book is about practical approaches, therefore the first question is to consider what practical issues drive a scientist or technologist to consider using a purification fusion in the first place. To this end, we will consider three areas of practical value: those where fusions are used to solve the technical problems encountered with the expression of mammalian proteins in microbial systems; those which can help where the production of bulk quantities of protein is required and the cost of goods becomes a major driving factor; and finally, those fusions which can be used in research to discover the biological function of proteins where the gene sequence is known, but the physiological function of a protein is unknown.

2 Protecting proteins from proteolysis

In the early days of genetic engineering, it was found that a mammalian protein, even when controlled by a strong bacterial promoter, did not accumulate in bacteria. Two factors seemed to be at work. The codon uses of amino acids were different in micro-organisms and in mammalian systems. A codon calling for a rare tRNA could cause a stutter in the expression system and termination of the protein's expression. Secondly, the bacteria seemed to be able to recognize foreign proteins and digest them, particularly if they were of low molecular weight. The practical problem faced by the genetic engineer was how to get the expression to a high level and to protect the protein from microbial proteases. These problems were not faced with endogenous bacterial proteins and so the desired protein was fused onto the N-terminus of a highly expressed bacterial protein. It worked, and the fused protein accumulated at very high levels. The reason why the protein accumulated was not because the proteases no longer digested the 'foreign' protein, but because the fused protein was expressed, precipitated and denatured, in an inclusion body. This dense protein agglomerate protected the protein from proteolysis.

The proteins were now successfully expressed and the inclusion body allowed significant purification from other bacterial proteins simply by lysis and centrifugation, but the proteins were denatured and linked to large unwanted protein fusions. However, protein chemists involved in protein sequencing had developed a number of chemical and enzymatic methods which allowed proteins to be selectively cleaved. It was realized that by incorporating these cleavage-sequences into the protein at specific points, a selective cleavage could be obtained which allowed the desired protein, or a slightly modified version of it, to be released (2) (*Table 1*). For certain peptides and proteins, refolding was easily achieved after the polypeptide fusion was removed but in other cases the protein remained inactive. However, the theoretical basis for protein denaturation and refolding had been sketched out in the 1960s where model proteins were used to study the effect of urea and guanidine on the denaturation of proteins. By trial and error, it was often possible to find conditions where significant yields of active, refolded proteins would be produced.

In order to express a small polypeptide ($< 10\,000$ Da), a polypeptide fusion

Table 1 Specific cleavage sequences[a]

Peptide	Method	Products
N-Met-C	Cyanogen bromide	N-Met and C
N-Asp-Pro-C	Acid	N-Asp and Pro-C
N-Lys/Arg-C	Trypsin	N-Lys/Arg and C
N-Glu/Asp-C	V-8 protease	N-Glu/Asp and C
N-Lys-Arg-C	Clostropain	N-Lys-Arg and C
N-(Lys)n/(Arg)m	Carboxypeptidase B	N and nLys and mArg
N-Asp-Asp-Lys-C	Enterokinase	N-Asp-Asp-Lys and C
Glu-Ala-Glu-C	Aminopeptidase 1	Glu-Ala-Glu and C
N-Ile-Glu-Gly-Arg-C	Factor Xa	N-Ile-Glu-Gly-Arg and C
N-Pro-X-Gly-Pro-C	Collagenase	N-Pro-X and Gly-Pro-C

[a] The amino acid sequences above (where X indicates any amino acid) have been used as fusion protein linkers to allow a specific cleavage which removes most or all of the fusion polypeptide from the desired protein. The peptides may be used with fusion polypeptides linked to either or both of their amino and carboxy termini. These are indicated as N- and C-polypeptides, respectively.

with a suitable cleavage site and expressing it in *E. coli* remains a practical and simple solution. Small peptides often do not contain a full complement of amino acids and this can greatly simplify finding a suitable cleavage site. In the examples below, small proteins were successfully expressed in *E. coli* by attaching the proteins to the N-termini of well-expressed bacterial proteins followed by a selective cleavage of the desired product using a commercially available protease.

2.1 Purification of fused APIII and release of protein using V8 protease

Atriopeptide III causes vasodilation and also acts as a diuretic. Both these effects can help in the treatment of high blood pressure. The peptide consists of only 24 amino acids and is unstable in normal *E. coli* expression systems due to rapid proteolysis. When expressed fused to a truncated bacterial protein RecA, the resulting protein is stable. A number of fusion linkers were investigated, factor Xa, thrombin, and V8. V8 protease is not a logical choice for this protein since it cleaves after glutamic or aspartic residues. Although APIII lacks glutamic acid, it does contain an aspartic acid. Nevertheless, a gene was constructed which expressed a fusion protein with an aspartic acid at the junction of the RecA protein and APIII, and unexpectedly, the desired cleavage was obtained (see *Figure 2*) (3). This approach illustrates the protection effects which can occur when a protein folds and shows the advantage of exploring a number of options when looking for a suitable cleavage site. After codons were optimized for expression in *E. coli*, the protein was expressed in a stable form, purified by centrifugation and chromatography, then dialysed against 50 mM NH_4HCO_3 for 6 h at 4°C, where oxidation of the APIII cysteines occurred. This oxidation was an important requirement to reduce the rate of cleavage of the internal aspartate. The protein digestion needs to be carefully monitored in order to prevent over-digestion which will yield $APIII_{1-9}$ and $APIII_{10-24}$.

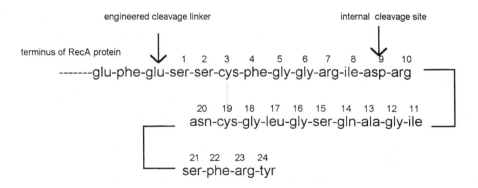

Figure 2 Primary sequence of APIII fused to the carboxy terminus of RecA using a glutamic acid linker for cleavage by V8 protease.

Protocol 1

Cleavage and analysis of oxidized RecA–Glu–APIII

Equipment and reagents

- 100 ml of approx. 1 mg/ml RecA-Glu-APIII in 50 mM ammonium bicarbonate clarified by centrifugation at 12 000 g for 10 min at 4°C in a screw-top polypropylene bottle
- 0.5 ml of V8 protease stock, 1 mg/ml
- Analytical HPLC fitted with a C-18 reverse-phase column, 10 μl loop injector, and UV detector set at 280 nm
- HPLC elution buffer: 100% acetonitrile, 0.05% trifluoroacetic acid

- HPLC equilibration buffer: 10% (v/v) acetonitrile/water and 0.05% trifluoroacetic acid
- Circulating water-bath at 37°C
- APIII standard and APIII cleavage standard prepared by reducing APIII with 10 mM dithiothreitol (DTT) and digesting with V8 protease for 4 h according to the conditions shown below

Method

1 Pre-incubate APIII (100 ml) and V8 protease stock to bring temperature to 37°C.

2 Add 0.1 ml of V8 protein, gently mix, and incubate.

3 Remove 0.5 ml sample and mix with 0.5 ml of HPLC equilibration buffer.

4 Inject 10 μl of sample and analyse using a 0–100% linear gradient.

5 Calculate the peak area ratio of APIII/RecA-Glu-APIII.

6 Stop the reaction when the ratio is equal to 95% by cooling the reaction mixture to 4°C and adding 10 ml of 0.1 M HCl.

2.2 Isolation of EGF from tryp E-fusion using trypsin

Human epidermal growth factor (EGF) is a 53 amino acid polypeptide hormone with ulcer and wound-healing properties. The protein had formerly been isolated in minute quantities from human urine, but it was only with the advent of genetic engineering that it was possible to produce it in sufficient quantities to allow its study as a potential medicine (4). If the native protein is expressed alone in *E. coli* it is rapidly degraded. However, if it is expressed as a fusion protein, it can be stabilized by forming inclusion bodies. The protein itself is resistant to trypsin digestion (even though it contains arginine), and so it was expressed as a N- and C-terminal fusion with arginine as the cleavage peptide for trypsin (see *Figure 3*).

The digestion of the protein with trypsin causes the release of acid. Therefore, in a dilute buffer, it is possible to follow the digestion of the protein by the amount of alkali added to the digestion mix in order to keep the pH at a fixed value. In this case, pH 8.0 was used as the optimal pH for the digestion. The use of an autotitrator allows the reaction to be followed continuously. The example is

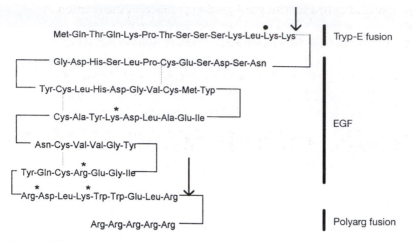

Figure 3 Trypsin cleavage of an N-terminal expression fusion and a C-terminal purification fusion to yield EGF. The desired trypsin cleavage sites to obtain EGF are shown. Additional cleavage sites within the protein are possible (*), but these are largely protected by the tight folding of EGF. And additional cleavage site leaving Lys–EGF as a major contaminant is shown (●).

at a large pilot plant scale with 5 g of protein. This was produced from a 100 litre fermentation, extracted and refolded in urea, purified by ion exchange chromatography, precipitated with ammonium sulfate, and dialysed against 10 mM acetic acid (5).

Protocol 2
Removal of N- and C-fusions using trypsin

Equipment and reagents

- Purified EGF-fusion protein (5 g) at 9.0 mg/ml extensively dialysed against 10 mM acetic acid
- Trypsin, 28 mg (*N*-tosyl-L-phenylalanine chloromethyl ketone treated) stock at 10.0 mg/ml

- Stopping buffer: 50 mM sodium acetate, 50 mM sodium chloride, adjusted to pH 3.6 with acetic acid
- Autotitrator loaded with 100 mM NaOH

Method

1 Gently stir the protein solution in a 1 litre glass beaker in a 37°C water-bath using an overhead stirrer.

2 Insert the probe from the autotitrator, set to pH 8, start recorder, and add trypsin. Continue digesting until no more acid is being produced then turn off titrator.

3 Add 500 ml of stopping buffer and adjust to pH 3.6 with acetic acid.

The protein is now contaminated with the digestion products of the reaction. Trypsin has cleaved the protein at the desired sites, leaving EGF and the expected fusion products, but as with any chemical reaction, undesirable side reactions have occurred. There have been some internal clips, some undigested protein, and some partially digested protein. Therefore, the protein must be purified from these cleavage products. In this case, the purification was achieved using a preparative scale high-performance ion exchange chromatography media.

Protocol 3

Purification of EGF from digestion products

Equipment and reagents

- Hydrophilic cation exchange HPLC column (500 ml, 200 × 55 mm dia) designed for protein separation (e.g. Biogel TSK-SP-5PW)
- Programmable preparative HPLC system capable of 100 ml/min and 1000 psi equipped with fraction collector and UV detector

- Trypsin-digested EGF at 2 mg/ml
- Buffer A: 500 mM sodium acetate, 50 mM sodium chloride pH 3.6
- Buffer B: 500 mM sodium acetate, 500 mM sodium chloride pH 3.6
- Rinse buffer: 0.1 M sodium hydroxide

Method

1   Equilibrate column with 1.8 litre of buffer A.
2   Load column with 500 ml of protein digest (1 g).
3   Wash with 1.8 litre buffer A.
4   Elute with 1.8 litre buffer B, collecting 50 ml fractions.
5   Strip with 450 ml rinse buffer.
6   Repeat process until all protein has been purified.

3   Engineering to improve purification processes

The next practical issue concerns reducing the cost of goods. This is a key issue for those proteins which are required in large quantities and at low cost, for example as biocatalysts, protein polymer materials, or foods. This is also likely to be an issue when the first phase of pharmaceutical rDNA derived proteins move into the generic market place. In this case, the cost of goods will become a more important driving force than 'first-to-market'. In all these cases, a careful use of fusions to improve the purification properties of the protein should be considered in order to optimize yields and reduce costs. Since most of these applications are in the commercial realm and are, therefore, trade secrets, few applications in this area have been published, but a number of approaches have been studied which illustrate the potential uses of this approach.

3.1 Fluidized bed IMAC for polyHis-tagged proteins

The His-Tag purification system exploits the affinity of histidine residues for metal cations. A chromatography support, derivatized with a chelating agent and charged with divalent cations such as Zn^{2+} or Ni^{2+} is used for immobilized metal affinity chromatography (IMAC) (6). A polyHistidine fusion has high affinity for these supports, and with optimization of the process, the fusion protein can be purified to a very high level in a single step. A metal chelate chromatography step is usually carried out using a conventional fixed bed, but at a process scale, fluidized bed chromatography offers many cost saving advantages. Use of IMAC in a fluidized bed chromatography system can allow initial cell debris removal (for example by microfiltration/centrifugation), ultrafiltration, and dialysis steps to be omitted. Overall, fluidized bed chromatography of a histidine-tagged protein can result in a substantial reduction in process costs (*Table 2*).

Table 2 Cost analysis of processes (data permission of Genzyme, Inc.)[a]

	Non-histidine tagged protein	Non-fluidized, histidine tagged protein	Fluidized histidine tagged protein
Cost/50 runs	$95 275	$77 235	$60 125
Yield/50 runs	25 mg	100 mg	1240 mg
Cost/mg protein	$3811	$772	$48

[a] The purification costs were compared using processes developed for the same protein without a histidine tag, with a 6-histidine tag using standard column chromatography, and with a 6-histidine tag in a fluidized bed mode. The initial process required eight separate steps, but with the tagged protein and fluidized bed chromatography, purification was achieved in three steps. It should be noted that if cost data is collected for comparable batch sizes, there is only a threefold saving in costs using a fluidized bed compared with conventional chromatography (7).

PolyHis-tagged proteins can bind to IMAC systems in high salt and denaturants. Multi-site attachment actually increases with increased buffer ionic strength (to 0.5 M and above) (8). IMAC will also operate in 6 M guanidine–HCl. This is a major advantage, since highly expressed fusion proteins in inclusion bodies can be readily separated in a denatured form early in the purification process, for subsequent refolding and further purification. However, the binding is weaker. With a polyHis-RecA protein in non-denaturing conditions a four-residue histidine tail and immobilized Zn^{2+} was optimal. In 6 M guanidine hydrochloride a 'stronger' metal ion (Cu^{2+} or Ni^{2+}) or a longer histidine tag on the protein (six residues) was required (9).

As well as the need to optimize the selection of the divalent complexing metals, the chromatography elution conditions need to be optimized. Imidazole buffers at low pH can cause loss of enzyme activity after elution. The use of EDTA as an eluent can also be detrimental since it could bind important metal cofactors. L-Histidine at pH 8.2 has been successfully used as an alternative eluent in the purification of the PutA protein from *S. typhimurium* where imidazole buffer caused a 22-fold reduction in enzyme activity (10).

Protocol 4

Fluidized bed chromatography of 6His-fused protein

The exact conditions will be protein-specific but the following will act as a guide to the process steps required for adsorption/elution metal chelate chromatography using a fluidized bed. The system has been scaled to handle the protein extracted from a 10 litre pilot fermentor in which *E. coli* is grown to a high cell density.

Equipment and reagents

- Prosep-Chelating chromatography adsorbent: a controlled-pore glass matrix derivatized with iminodiacetic acid covalently attached by a stable hydrophilic spacer (Bioprocessing Ltd.)
- Fluidized bed column, for example the Amicon Moduline, containing approximately 1 litre packed bed of chromatography adsorbent. A standard chromatography column can also be converted into a fluidized bed system by inserting a 50 μm inlet mesh.
- 30 mM phosphate pH 7.5

- *E. coli* extract containing a 6His-fused protein, at pH 8.0 (typically 10 litre)
- 0.1 M $NiSO_4$ pH 3.3
- 0.25 M NaCl in distilled water
- Buffer A: 50 mM imidazole, 30 mM phosphate pH 7.5
- Buffer B: 1.0 M imidazole, 30 mM phosphate pH 7.5
- 0.4 M EDTA
- 0.3% HCl pH 1.5

Method

1   After the cleaning procedure, fluidize the bed with 5 litres of 0.25 M NaCl in ascending flow.

2   Charge chromatography medium with metal using 1.5 litres of 0.1 M $NiSO_4$ pH 3.3.

3   Wash with 5 litres of 0.25 M NaCl in distilled water to remove free metal, then 5 litres of 30 mM phosphate buffer pH 7.5 to remove loosely bound metal, then equilibrate column in 10 litres of the same buffer.

4   Load the *E. coli* extract (4–15 litres) and wash to remove unbound protein with 5 litres of 30 mM phosphate pH 7.5, then with 5 litres of buffer A.

5   At this point the bed is allowed to settle and protein is eluted from the packed bed using downward flow using a 10 litre linear gradient of buffers A and B.

6   Clean the column with buffer A, strip the metal using 5.0 litre 0.4 M EDTA, and regenerate with 5 litres of 0.3% HCl pH 1.5.

7   Repeat the cycle.

Magnetizable particle separation technology is also available for IMAC. MagIMAC (Scigen Ltd.) is made by covalently linking magnetic agarose beads with iminodiacetic acid and this reacts with metal ions as described above. The difference is that the paramagnetic chromatography matrix, with its bound protein fusion, can easily be removed from the crude feedstock after adsorption by applying a magnet to the wall of the vessel. After a wash step to remove debris and contaminants, the fusion protein is eluted from the matrix. This method is highly suitable for multiple, exploratory batch purifications from mixtures containing large amounts of cellular debris. The general method has been applied successfully to the isolation of a polyHistidine-tailed T4 lysozyme from an *E. coli* extract, using Cu^{2+} as the counter ion with iminodiacetic acid-derived Sepharose (11).

3.2 Engineering metal affinity sites in somatotropins

Somatotropins are protein hormones involved in the regulation of the growth of animals. Human growth hormone is used to treat dwarfism in children, and bovine growth hormone to increase lactation in cows. Both of these are commercial products and a considerable amount of information is available about these proteins. The details of metal chelation are described elsewhere in *Protein Purification Techniques* and the structural considerations required to design the protein have already been described (2). The method used was to engineer two surface histidines which are held in a rigid orientation to each other. From the primary structure of the proteins and using β-sheet, α-helical prediction algorithms, somatotropins were predicted to contain a number of α-helical structures. Using predictions of hydropathy, the exposed rather than buried sections of the α-helical sections were identified. All three of the protein's histidines were in these regions and two of the three were in a suitable orientation to the solvent (His 19 and His 169). For an α-helix, structural considerations require His-X-X-X-His to provide the correct orientation of the second His in order to provide a metal chelation site. Therefore, the gene was re-engineered to code for a His 15 in place of the natural Leu 15. As predicted, the expressed protein had the required increased binding properties for metal chelation chromatography.

Protocol 5

Metal affinity chromatography

Equipment and reagents

- Copper metal affinity chromatography media: iminodiacetic acid based
- Equilibration buffer: 50 mM NaH_2PO_4 pH 7.0
- Column with gel bed size of 10 mm × 130 mm

- Chromatography pumping system and fraction collector capable of 2 ml/min
- Extract from 100 g of bacterial paste containing engineered protein equilibrated in 4 mM urea/borate pH 9
- N-α-acetylhistidine

Method

1 Pack column and equilibrate with buffer.
2 Apply protein solution at 1 ml/min, wash with 4 mM borate buffer pH 9.0.
3 Wash with 40 ml extraction buffer.
4 Apply N-acetylhistidine gradient (1–97 mM) 280 ml at 2 ml/min.

3.3 Engineering proteins for polyelectrolyte precipitation

Early in a purification, precipitation can be used to concentrate impure and dilute proteins. By carefully controlling the levels of precipitating agents, it is possible to cause a selective precipitation of the desired protein while leaving the impurities in solution. This process has the potential for large scale and low cost, requiring only a stirred tank and a centrifuge (12). It is used to process soya proteins on the multi-ton scale. The use of charged purification fusions to enhance the selectivity of polyelectrolyte precipitation methods has been the subject of considerable study. Polyaspartic acid and polyarginine fusions have both been constructed for β-galactosidase. Using poly(ethyleneimine) for the polyaspartic acid fusion and poly(acrylic acid) as the precipitants, the expected change in precipitation characteristics occurred, with the proteins being precipitated at considerably lower concentrations than the unmodified control. With crude samples, precipitation of the polyaspartic acid fusion was complicated by the presence of nucleic acids which were preferentially precipitated but the polyarginine-fused proteins still retained their enhanced susceptibility to precipitation (13). The procedure below uses commonly available proteins as an illustration. The use of a particular fusion protein could change the optimal polyelectrolyte dosage but the procedure would otherwise be the same. As long as the protein and polyelectrolyte complex well, the absolute concentrations are not very important to yield, only the ratio of the two. Higher concentrations will generally lead to larger, more easily settled precipitate particles.

Protocol 6

Precipitation of egg white with PAA (polyacrylic acid)

Notes

The precipitations are done at room temperature (23 °C) in phosphate buffer (10 mM, pH to 7.33 ± 0.02) where egg white lysozyme (EWL) is positively charged and bovine serum albumin (BSA) is negatively charged. Buffer is not necessary, but makes it easier to control the pH as the protein and polyelectrolyte each also function as acid/base/buffering components. Elevated salt concentrations will weaken the complexation. For this procedure final concentrations of individual proteins, after combination with the precipitant, are 0.2 mg/ml. Larger volumes are even easier to work with and allow for more rapid mixing of components which prevents over-precipitation. The volume differences in the samples caused by the addition of the polyelectrolyte can be compensated for by the addition of buffer. Protein removal can be followed by assay of the supernatant protein content determined by A_{280} or a colorimetric soluble protein assay such as the bicinchonic acid (BCA) protein assay (Pierce Chemicals). The enhanced protocol of this assay gives less variability than the standard assay. For A_{280} use the extinction coefficient values of 5.8 for BSA and 26.4 for EWL. Protein removal is generally linear with added dosage of polyelectrolyte until near the optimum removal of protein. Beyond this dosage yield of precipitate may increase due to formation of soluble complexes of protein and polyelectrolyte.

Protocol 6 continued

Reagents

- Proteins EWL and BSA (Sigma Chemical Co.)
- PAA (150 000 Da) is a 25 wt% solution (Polysciences) and polyphosphate Glass H (5000 Da) is in powder form (FMC Corp.) (less satisfactory results can be expected with lower molecular weight PAA)

- Protein stock solutions (10 mg/ml) are filtered (0.45 μm) and dialysed against the phosphate buffer (10 mM, pH 7.33 ± 0.02)
- Polyelectrolyte stock solutions: 1 mg/ml in phosphate buffer (10 mM, pH 7.33 ± 0.02)

Method (14)

1 Dispense 0.02 ml of 10 mg/ml stock protein (0.2 mg) into a series of 1.5 ml micro-centrifuge vials.

2 Add polyelectrolyte stock solution in dosages ranging from the underdosed to over-dosed (0–125 mg PAA per g EWL, 0–300 mg Glass H per g EWL, and 0–1500 mg PEI per g BSA) to the protein solution.

3 Add buffer to make up the volume to 1 ml (final protein concentration of 0.2 mg/ml).

4 Mix the samples thoroughly with the polyelectrolyte by vortexing for 5 min.

5 Separate the precipitated material by centrifugation at 4°C and 6400 g for 15 min (for these volumes: Model 235B microcentrifuge, Fisher Scientific. Redissolution of the precipitate is easier if the mildest centrifugation possible is used).

6 Carefully pipette the solution away from the pellet and redissolve in buffer.

7 Measure the protein concentration in the supernatant.

4 Fusions of generic application

A researcher may be faced with the purely practical problem of isolating a protein in a sufficiently pure state in order to study it and understand its function. Here the problem to be solved is one of speed and convenience rather than cost of goods. Until recently, engineering the protein would only be considered if it was proven to have some useful properties. However, with the advent of genomics, the function of a protein is either unknown or, being based on structural analogies with proteins of known function, merely hypothetical. Therefore, in order to study the protein's real function, it will need to be expressed and purified. The question is how to maximize the chances of obtaining the expressed protein, not only purified but also in its natural biologically active form. A good solution is to use a purification fusion which does not interfere with the protein's activity, either because it can be easily removed or does not affect the protein's folding. The expression system used must also be as close as possible to the genetic source of the material (mammalian cell lines for human proteins, plant cell lines for plant proteins, etc.) so that the chances for the correct folding of the protein are maximized.

Table 3 Options for protein fusions[a]

General type	Size	C- or N-terminal	Separation method(s)
1. Enzymes			
beta-gal	116 kDa	N, C	IS, IA
GST	26 kDa	N	IS (GSH)
CAT	24 kDa	N	IS
Tryp E	27 kDa	N	HIC
2. Polypeptide-binding proteins			
SPA	14–31 kDa	N	I-IgG
SPG	28 kDa	C	I-albumin
3. Carbohydrate-binding domains			
MBP	40 kDa	N	IS
SBD	119 aa	C	IS
CBD (CenA)	111 aa	N	IS
CBD (Cex)	128 aa	C	IS
4. Biotin-binding domain	8 kDa	N	I-avidin
5. Antigenic epitopes			
RecA	144 aa	C	IA
Flag	8 aa	N	IA
6. Charged amino acids			
Poly(Arg)	5–15 aa	C	IEC, pptn
Poly(Asp)	5–16 aa	C	IEC, pptn, RME
Glutamate	1 aa	N	IEC
7. Poly(His) tails	1–9 aa	N, C	IMAC
8. Other poly(amino acid) tails			
Poly(Phe)	11 aa	N	Phenyl–Sepharose
Poly(Cys)	4 aa	N	Thiopropyl–Sepharose

[a] Abbreviations used: aa, amino acids; beta-gal, β-galactosidase; CAT, chloramphenicol acetyltransferase; CBD, cellulose-binding domain; GSH, reduced glutathione; GST, glutathione transferase; HIC, hydrophobic interaction chromatography; IA, immobilized antibody; IEC, ion exchange chromatography; I-IgG, immobilized immunoglobulin G; IMAC, immobilized metal affinity chromatography; IS, immobilized substrate; MBP, maltose-binding protein; pptn, selective precipitation; RME, reverse-micellar extraction; SBD, starch-binding domain; SPA, staphylococcal protein A; SPG, streptococcal protein G.

The list of tail types which can be used to purify proteins has been recently reviewed (12) and is quite extensive (*Table 3*). Any of the smaller tags are likely to be the most useful. However, the advantages offered by purification fusions are now recognized by suppliers, who offer kits and reagents based on specific fusion technologies. These have taken the technology of protein expression and purification to an extremely high level of sophistication. If the target protein is suitable for application of fusion techniques, these kits enable the user to clone the protein, express it in a suitable host, and purify it by an affinity method. However, since there is no single solution to the problems encountered when attempting to isolate proteins produced by recombinant DNA technology, the best strategy will be to try several methods and find the one which works best for a given system.

4.1 PolyHistidine

Sigma offers a kit for producing and purifying proteins with a polyHistidine tag (a sequence of six histidines). This tag has a high affinity for divalent metal ions with which it can form a chelate. Since the poly6His does not occupy all the co-ordination sites of the bound metal, the metal ion can be simultaneously complexed with the polyHis-tagged fusion protein and the complex chromato-graphed using IMAC. Nitrilotriacetic acid (NTA) is immobilized on agarose and the NTA-agarose pre-loaded (typically) with nickel (Ni^{2+}). The NTA agarose is then used to bind to polyHistidine-tagged fusion proteins. PolyHistidine has a well-characterized affinity for nickel–NTA-agarose, and can be eluted by a num-ber of methods, e.g. imidazole buffer at 0.1–0.2 M concentration will effectively compete for resin-binding sites, and lowering the pH to just below 6.0 proto-nates the histidine side chains and disrupts the binding. A gradient is frequently used. The high affinity of nickel for polyHis makes it effective in the purification of polyHistidine-tagged fusion proteins from a typical expression system. Sigma offers five more metal–NTA-agaroses, giving scope for optimizing separations of fusions of varying chain lengths and size of target molecule. PolyHistidine fusion and IMAC kits are also supplied by Amersham Pharmacia Biotech and Novagen.

4.2 FLAG (2)

Developed by Immunex and Kodak and now marketed as a kit by Sigma, this is a proprietary system designed to facilitate expression and purification of recombi-nant proteins in *E. coli* and yeasts. A complete system includes vectors and strong promoters to drive the synthesis of recombinant proteins and a FLAG peptide is encoded at either the C- or N-terminus of the target protein. FLAG is an octa-peptide designed for easy detection and purification of the fusion protein. It has the sequence N-Asp-Tyr-Lys-Asp-Asp-Asp-Asp-Lys-C. The FLAG peptide is 'recog-nized' by an immobilized anti-FLAG antibody and is the binding ligand for protein purification on the anti-FLAG affinity gel. Antibody/antigen binding is calcium dependent. This N-terminal peptide also contains a highly specific enterokinase cleavage site. Therefore, this enzyme may be used to selectively remove the fusion after purification. Kits for expression in *E. coli* are supplied in two con-figurations, allowing N-terminal or C-terminal expression of the FLAG peptide. The yeast expression kit is configured for N-terminal expression only. The *E. coli* amino terminal FLAG expression kit contains the following:

- expression vectors
- sequencing primers
- anti-FLAG monoclonal antibodies
- affinity gel FLAG peptide
- control plasmid and protein
- enterokinase

Reagents are also available separately, for example a purified murine IgG2b monoclonal antibody, anti-FLAG M1, isolated from murine ascites fluid; this anti-body binds to proteins with a FLAG marker at the free N-terminus. Other anti-bodies are also available, for use according to species and location of the fusion sequence.

4.3 Recombinant phage antibody system

A kit, called the Recombinant Phage Antibody System (RPAS) Purification Module is supplied by Amersham Pharmacia Biotech. This is designed for one-step affinity purification of soluble single-chain fragment variable (ScFv) antibodies that carry a C-terminal 13 amino acid peptide tag (E-Tag). Purification is based on Pharmacia's anti-E-Tag Sepharose high-performance medium. Genes for the heavy and light chains are isolated from mouse hybridoma or spleen cells and assembled into an ScFv gene using the RPAS mouse ScFv module. Once constructed, the ScFv gene is inserted into an expression vector for production of phage-displayed or soluble ScFv antibodies that contain an E-Tag. The E-Tag is recognized by an anti-E-Tag monoclonal antibody that has been covalently coupled to the NHS-activated Sepharose for affinity purification of ScFv antibodies. The kit comprises the following:

• a pre-packed anti-E-Tag affinity column, binding capacity ca. 0.7 mg ScFv at saturation

• buffers for 10 purifications (the anti-Tag column can be used at least 20 times)

• column adapters

4.4 Fusion tag detection and purification: S·Tag system

This protein tagging system is based on the interaction of the 15 amino acid S·Tag peptide with a 104 amino acid long S-protein derived from pancreatic ribonuclease A. It is supplied by Novagen. An S-protein : S-peptide complex (known as ribonuclease S) has enzymatic activity, whereas neither component has activity by itself. These characteristics make the S·Tag peptide a convenient tag for recombinant proteins, since the fusion protein can be detected and purified using the S-protein. The reconstituted enzymatic activity caused by the S·Tag-peptide : S-protein interaction allows a sensitive and quantitative measurement of the fusion protein using a simple assay. Kits are available for detection of S·Tag fusion proteins and include the S·Tag rapid assay kit for quantitative measurement and the S·Tag Western blot kit for detection of fusion proteins on blots. The S·Tag purification kit takes advantage of the S·Tag-peptide : S-protein interaction for rapid affinity purification of fusion proteins on a small to medium scale, under native conditions.

4.5 T7·Tag affinity purification kit

Novagen also supplies a T7·Tag affinity purification kit designed for rapid immunoaffinity purification of target proteins that carry the 11 amino acid T7·Tag sequence (i.e. the initial 11 amino acid of the T7 gene 10 protein). Purification is based on binding target proteins to anti-T7·Tag monoclonal antibody which is covalently coupled to cross-linked agarose beads. Bound protein is usually eluted at pH 2.2 but neutralization buffer is included to limit protein exposure to low pH. The capacity varies between different target proteins, but the beads are standardized to bind a minimum of 300 µg T7·Tag-galactosidase per ml of settled resin. The beads can be used in either batch or column modes and can be recycled a minimum of five times without loss of binding activity. The kit contains T7·Tag antibody agarose, buffers, and column.

4.6 Fusions based on maltose-binding protein

New England Biolabs supplies a kit based on a fusion of the protein of interest to maltose-binding protein (MBP, encoded by a *malE* gene). This fusion can be expressed in large amounts using a strong promoter and translation initiation signals of MBP. A one-step affinity purification for MBP is used to purify the fusion protein. The system contains pMAL vectors, which include convenient cloning sites for the gene of interest. Fusion protein yields of up to 100 mg from a litre of culture are claimed for this system. The vector also includes a sequence coding for the recognition sites of specific proteases, e.g. factor Xa. This sequence allows cleavage of the target protein after purification on the amylose resin affinity column.

4.7 Intein-mediated purification with an affinity chitin-binding tag (15)

New England Biolabs supplies a kit based on mechanistic studies of protein splicing. The IMPACT I System uses a protein splicing element, an intein, from *Saccharomyces cerevisiae VMA1* gene. The intein has been modified to self-cleave at its N-terminus at low temperatures in the presence of thiols such as, 2-mercaptoethanol or cysteine. The gene encoding the target protein is inserted into a multiple cloning site (MCS) of a pCYB vector to create a fusion between the C-terminus of the target gene and the N-terminus of the gene encoding the intein (*Figure 4*). The DNA encoding a small 5 kDa chitin-binding domain (ChBD) from *Bacillus circulans* has been added to the C-terminus of the intein for affinity purification of the three-part fusion. Expression of the fusion construct is controlled by a Ptac promotor. When crude extracts of cells from an inducible *E. coli* expression system are passed through a chitin column, the fusion protein binds to the chitin column while all other contaminants are washed through the column. The fusion is then induced to undergo an intein-mediated self-cleavage on the column by overnight incubation at 4°C in the presence of DTT or 2-mercaptoethanol. The target protein is released while the intein-chitin binding domain fusion partner remains bound to the column (*Figure 4*). This system presents many advantages, such as simultaneous purification and cleavage all at low temperature.

5 Concluding remarks

Since the early 1980s, purification fusions have developed from being a 'quick fix' for the proteolysis of mammalian protein expressed in bacteria, to become highly sophisticated generic purification systems. These purification fusions are the first examples of the art of protein chemistry turned into a science of pro-

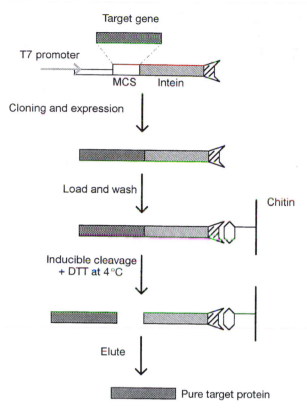

Figure 4 Principle of the IMPACT T7 affinity fusion.

tein engineering. The technology is versatile over scale and applicable not only to low cost, large scale protein manufacture but to also meet the needs of the researcher for rapid and simple isolation methods. However, the complexities of protein production and folding still means that there is a considerable amount of art involved in protein isolation. Consequently, it is advisable to screen a number of fusion expression systems in order to find the best fit for the protein and its expression system. It is hard to see further dramatic changes in the technology over the next few years. Instead, there will be a consolidation and application of the technology to a wider range of proteins and expression systems, particularly at very large scale, where plants are most likely to become the expression system of choice.

References

1. Brewer, S. J. and Sassenfeld, H. M. (1990). In *Protein purification applications: a practical approach* (ed. E. L. V. Harris and S. Angel), pp. 91–112. IRL Press, Oxford, UK.
2. Brewer, S. J., Haymore, B. L., Hopp, T. P., and Sassenfeld, H. M. (1991). In *Purification and analysis of recombinant proteins* (ed. R. Seetharam and S. K. Sharma), pp. 239–66. Marcel Dekker Inc., New York, USA.
3. Mai, M. S., Bittner, M. L., and Braford, S. R. (1992). US Patent 5087564.

4. Brewer, S. J., Dickerson, C. H., Ewbank, J., and Fallon, A. (1986). *J. Chromatogr.*, **362**, 443.

5. Smith, J. C., Derbyshire, R. B., Cook, E., Dunthorne, L., Viney, J., Brewer, S. J., *et al.* (1984). *Gene*, **32**, 321.

6. Yip, T. T. and Hutchens, T. W. (1994). *Mol. Biotechnol.*, **1**, 151.

7. Bigelow, R. (1995). Paper presented at *ISPE,* Boston, JEP, Session I-F.

8. Wei Jiang and Hearn, M. T. W. (1996). *Anal. Biochem.*, **242**, 45.

9. Report. (1995). *Downstream*, Newsletter from Amersham Pharmacia Biotech No. 20.

10. Gort, S. and Maloy, S. (1998). Elsevier Trends Journals Technical Tips Online No. T01389. http://tto.trends.com.

11. O'Brien, S. M., Sloane, R. P., Thomas, O. R. T., and Dunnill, P. (1997). *J. Biotechnol.*, **54** (1), 53.

12. Niederauer, M. Q. and Glatz, C. E. (1992). *Adv. Biochem. Eng. Biotechnol.*, **47**, 159.

13. Ford, C. F., Suominen, I., and Glatz, C. E. (1991). *Protein Expression Purification*, **2**, 95.

14. Sternberg, M. and Hershberger, D. (1974). *Biochim. Biophys. Acta*, **342**, 195.

15. Chong, S. (1996). *J. Biol. Chem.*, **271**, 22159.

Chapter 2
Initial purification of inclusion bodies

Bernard N. Violand

Monsanto Company, 700 Chesterfield Parkway North, Mail Zone AA YI, St Louis, Missouri, MO 63198, USA.

1 Introduction

Recombinant DNA technology has made possible the production of many proteins which were previously available in only limited quantities from natural sources. This technology has resulted in the generation of many therapeutically important protein products and large numbers of other commercially significant proteins. High level expression of both heterologous and homologous proteins frequently results in the formation of dense amorphous insoluble aggregates, named inclusion bodies, in the host cytoplasm. This chapter discusses some of the properties of inclusion bodies and also describes the most frequently used methods for preparing them. Because the majority of this data has been generated in *E. coli*, this chapter will concentrate on production of inclusion bodies from this host cell.

2 Mechanism of inclusion body formation

High level expression of recombinant proteins in *E. coli* can result in 40–50% of the total cell protein being the expressed protein. At these high expression levels, inclusion bodies frequently form in the cytoplasm with more than one inclusion body sometimes being present per cell. The formation of inclusion bodies during heterologous protein expression was first reported for production of proinsulin in *E. coli* (1) and has subsequently been reported numerous times. Despite the large number of proteins which have been expressed as inclusion bodies and numerous studies on their production, the exact molecular mechanism(s) by which they form is still unclear.

The majority of investigations on inclusion body formation have utilized *E. coli* as the host organism, however, other hosts can also form inclusion bodies during high level expression. These include *Saccharomyces* sp. (2), *Bacillus* sp. (3), and insect cells (4). The characteristics of the protein, the rate of expression, the level of protein expression, growth temperature, and media composition have sometimes been shown to have an effect on the propensity for formation of inclusion bodies versus soluble protein. However, each of these parameters may not affect all proteins to the same extent.

The expression of the restriction enzyme *EcoRI* demonstrates how the expression levels affects the propensity for a protein to forming inclusion bodies. At low levels of expression this enzyme is present as a soluble protein, however, at high expression levels it is produced in inclusion bodies (5, 6). A study on a limited number of proteins demonstrated that a lower growth temperature has a profound effect on the amount of protein present as a soluble form in the cytoplasm of *E. coli* for several proteins (7). This correlation of a lower growth temperature and more soluble protein has been subsequently shown to exist for numerous other proteins. Another study of 81 proteins analysed the correlation between six different protein properties and the likelihood for these proteins to form inclusion bodies (8). The protein properties which showed very little correlation to inclusion body formation included number of cysteines, number of prolines, hydrophobicity, and size of the protein. The parameters with the best correlations to inclusion body formation were charge average and turn forming residue fraction, however, these two properties are not expected to always be predictive of whether a protein will form inclusion bodies.

Another factor affecting inclusion body formation is that the cytoplasm of *E. coli* is maintained in a reducing environment which generally prevents the formation of disulfide bonds for expressed proteins. Therefore, disulfide bond formation will occur slowly or not at all in the cytoplasm and this may explain why some proteins which require disulfide bonds aggregate and form inclusion bodies. This cannot explain inclusion body formation in many instances, however, since proteins without disulfide bonds can also form inclusion bodies. In summary, there presently is no one parameter which is predictive for whether a particular protein under specified conditions will form inclusion bodies.

Recent investigations on the mechanism of inclusion body formation have resulted in a clearer understanding of this phenomena. These results indicate that the most likely explanation for inclusion body formation is aggregation of partially folded intermediates. Studies of interleukin-1β (9, 10), P22 tailspike (11) and coat proteins (12), and β-lactamase (13) have shown that partially folded species have significant secondary structure and associate through specific interactions with each other. *In vitro* studies on these same proteins are consistent with the theory of unstable transient folding intermediates being the major explanation for inclusion body formation.

3 Isolation of inclusion bodies from cell homogenates

Recovery of inclusion bodies from cell homogenates is usually accomplished by either centrifugation or filtration. Because of the dense nature of inclusion bodies, they can be differentially centrifuged from other cell debris. Cell homogenates will contain many constituents, including soluble proteins, insoluble cell debris, membranes, nucleic acids, other cell components as well as the inclusion bodies. Inclusion bodies are most commonly recovered from these homogenates using lab scale centrifuges. Centrifugation for 10–30 min at 10 000–20 000 g is usually sufficient to sediment the inclusion bodies from the soluble material. However, the exact conditions for centrifugation should be evaluated for each protein since the size of the inclusion bodies can vary depending on the protein and expression level and their size will affect their rate of sedimentation. For large scale recovery of inclusion bodies other types of centrifuges such as the continuous disc-stack type have been utilized (14). Filtration has also been used to recover inclusion bodies, however, it does not appear to offer any unique advantages over centrifugation (15, 16). *Protocol 1* describes a general procedure used for isolation of inclusion bodies from *E. coli* homogenates.

Protocol 1

Isolation of inclusion bodies

Equipment and reagents

- Lab scale centrifuge: Sorvall RC5C or equivalent
- Manton-Gaulin 15 M homogenizer with knife-edged valve
- Buffer A: 50 mM Tris–HCl, 5 mM EDTA pH 7.5
- Tekmar Ultra-turrax Model TR5T

Method

1 Suspend the cell paste in buffer A at 10 ml of buffer per 1 g of paste.

2 Homogenize thoroughly on ice with the Ultra-turrax maintaining a temperature < 10 °C.

3 Pass the cell suspension through the Manton-Gaulin at 10 000–12 000 psi.

4 Collect the homogenate into a container cooled on ice.

5 Pass 50 ml of cold water through the Manton-Gaulin to rinse out the remaining cell homogenate. Add this to the previously collected homogenate.

6 Cool down homogenate on ice until 4–6 °C.

7 Repeat steps 3–5.

8 Collect the inclusion bodies by centrifuging in a GSA rotor or equivalent at 10 000 g for 30 min at 4 °C. Pour off the supernatant, being careful not to disrupt the inclusion body pellet. If there is obvious cell debris on top of the inclusion bodies, scrape off this layer using a spatula. The inclusion bodies are usually light brown to slightly red coloured, whereas the non-desirable upper layer may be lighter in colour.

9 Add cold water to the inclusion bodies at half the original volume of buffer used for the homogenization.

10 Homogenize thoroughly using the Ultra-turrax.

Protocol 1 continued

11 Pass the cell suspension through the Manton-Gaulin once more at 10 000–12 000 psi and rinse out as before with 50 ml of cold water.

12 Collect the inclusion bodies as described in step 8.

13 Repeat steps 9 and 10.

14 Collect the inclusion bodies by centrifugation as described in step 8.

15 If considerable colour or cloudiness is still present in the supernatant after centrifugation, repeat the homogenization with the Ultra-turrax and recollect the inclusion bodies.

4 Washing of inclusion bodies

The goal of washing inclusion bodies is to improve the ease with which the desired protein can be subsequently refolded and purified. Washing may remove proteases and thus prevent instability of the protein during subsequent purification steps. Washing may also be used to remove DNA, lipids, and soluble cellular proteins and components before solubilization of the inclusion bodies. The washing steps may or may not be necessary depending on the subsequent purification steps and the purity of the initial inclusion bodies. We have observed instances where large amounts of DNA or other contaminants may be present in the inclusion bodies and a washing step significantly improves the subsequent refolding yield and/or stability of the protein against proteolysis. There are also cases where contaminating cell debris and soluble components interfere with subsequent chromatography steps by binding tightly to the column resin. With large scale production of proteins, such as bovine somatotropin, including a wash step may be cost beneficial since this simple step may improve the reproducibility and lifetime of expensive chromatography resins used in downstream processing. For proteins purified on a smaller scale and for which the cost of goods is not a major issue, the washing step may not be necessary from a practical point of view. Another advantage of washing the inclusion bodies is that it may reduce the level of pyrogens (mostly cell wall components) which may need to be removed prior to *in vivo* testing of the product.

A recent study on the effect of contaminants such as DNA, phospholipids, lipopolysaccharides, and other proteins on the renaturation of hen egg white lysozyme demonstrated that there was very little effect except at high concentrations of added contaminating proteins (17). Addition of certain phospholipids actually improved the yield from the refolding step. This was postulated to occur because they may act as detergents and help segregate the monomers from each other during refolding.

In general it is recommended to at least wash the inclusion bodies several times with water or alkaline buffers to prepare a reasonably pure inclusion body. The necessity of this wash will be dependent on how efficient the cell breakage and homogenization has been and the desired purpose of purifying the protein.

Since unbroken cells will co-sediment with the inclusion bodies it is important to visually inspect by phase contrast microscopy to ensure that the homogenate contains no intact cells. Generally, two to three passes of the cells at high pressure, $> 10\,000$ psi in a Manton-Gaulin, is sufficient to efficiently break $> 95\%$ of the cells. Because the efficiency of lysis and homogenization is dependent on operating pressure, higher pressures should be utilized whenever possible. For small sample volumes, sonication may be the preferred method for cell breakage, however, it is difficult to scale-up because of heat generation. Numerous short pulses (5–20 sec) intermittent with cooling of the sample can be used to maintain the sample at the desired temperature.

There are many examples of different wash solutions used to remove undesirable contaminants from inclusion bodies. For example, adding lysozyme, EDTA, and deoxycholate to the entire cell homogenate of bovine somatotropin inclusion bodies removed the majority of the nucleic acids, lipopolysaccharides, and phospholipids without solubilizing the bovine somatotropin (18). Generally, the inclusion bodies are first isolated by centrifugation and then washed several times. Triton X-100 (19–21), sucrose (22, 23), urea (24, 25), deoxycholate (26), and octyl glucoside (27) have all been used successfully for removing contaminants from inclusion bodies. The pH at which the washes are performed can be critical because at higher pHs, inclusion bodies will generally be more easily solubilized. Low pH washes (pHs of 3–6) of E. coli inclusion bodies are usually ineffective since the majority of the E. coli contaminating proteins are insoluble in this pH range. Since each protein in inclusion bodies has its own unique solubility characteristics, there is no one wash buffer which will be applicable to all inclusion bodies.

Some of the proteins shown to be been present in inclusion bodies include the outer membrane proteins OmpF, OmpC, and OmpA (28), the four subunits of RNA polymerase (29), the large ribosomal subunit protein L13 (29), elongation factor Tu (30), kanamycin phosphotransferase (31), and two small heat shock proteins, IbpA and IbpB (32). The question of whether these proteins are an integral part of inclusion bodies, adsorbed onto the surfaces of the protein inclusion bodies before or after cell lysis or are just part of the precipitated material which sediments with the inclusion bodies is not fully understood. However, most data indicates that these contaminants are not specifically targeted for inclusion bodies but they are present through co-sedimentation or are weakly adsorbed to the surface of the inclusion bodies. Membrane fragments present in cytoplasmic inclusion bodies of β-lactamase could be removed (27) using detergent washes and a sucrose step gradient. To prepare very pure inclusion bodies for subsequent characterization studies, a sucrose step gradient may be preferred instead of centrifugation or filtration.

We, as well as others, have determined that effective removal of the membrane components may be crucial for efficient purification since some proteases are membrane bound. The protease OmpT, which cleaves at dibasic residues, has been an especially prevalent protease in inclusion bodies. We have observed the cleavage of the dibasic Arg^{182}–Arg^{183} bond in bovine somatotropin if the inclusion bodies are not thoroughly washed and have determined that OmpT was responsible. Construction of an OmpT minus strain was shown to remove the proteolytic activity which had been cleaving β-galactosidase (33). Proteolysis of creatine kinase and bovine pancreatic trypsin inhibitor expressed in *E. coli* was a problem until a Triton X-100 washing step of the inclusion bodies was used (34). This washing most likely removed cell membranes containing proteases demonstrating that efficient washing may be critical to stabilization of the protein for subsequent purification.

Inclusion of protease inhibitors in the wash buffer may be necessary to prevent undesired proteolysis. The choice of inhibitors will be dependent on the target protease which may not be known. General serine protease inhibitors such as PMSF are frequently added to wash buffers and to the buffer in which the cells are homogenized. Our experience has shown that an efficient washing procedure is usually preferable to trying to inhibit a protease of unknown origin.

Some of the major factors which can affect the efficiency of washing inclusion bodies are:

- inclusion body protein solubility
- pH
- ionic strength
- temperature
- time of exposure
- efficiency of homogenization
- ratio of buffer to protein
- concentration of additive (urea, detergent)

Utilization of a high pH buffer (Protocol 2) will often be sufficient for preparing high quality inclusion bodies. Generally, a higher pH is more effective at removing contaminants as long as the pH does not solubilize the inclusion bodies nor chemically modify the protein of interest. *Protocols 2* and *3* describe two processes for washing inclusion bodies.

Protocol 2

Washing of inclusion bodies at basic pH

Equipment and reagents
- Lab scale centrifuge: Sorvall RC5C or equivalent
- Wash buffer: 10 mM $NaHCO_3$, 1 mM EDTA pH 10.0
- Tekmar homogenizer, Model TR5T or equivalent

Method
1   Collect the inclusion bodies by centrifuging at 10 000 g for 20 min at 4 °C.

Protocol 2 continued

2 Pour off the supernatant being careful not to disrupt the inclusion body pellet. If there is obvious cell debris on top of the inclusion bodies, you may want to scrape of this layer using a spatula. The inclusion bodies are usually light brown to slightly red coloured, whereas the non-desirable upper layer may be slightly lighter in colour.

3 Add an amount of wash buffer equal to 20 ml per gram of inclusion body.

4 Place the tube or bottle containing the inclusion bodies on ice and resuspend with the Tekmar homogenizer. This will usually take several minutes depending on the volume of inclusion bodies and buffer. Sonication using 10–20 sec bursts can also be used to resuspend the inclusion bodies. Be sure to monitor the temperature during sonication to ensure it does not exceed 10°C and keep the sample on ice during the sonication.

5 Recentrifuge as in step 1 and repeat steps 2–5 until the supernatant is clear and colourless.

6 Ensure that the protein of interest in the inclusion bodies is not being solubilized by this treatment by SDS gel electrophoresis or another analytical method suitable for this protein. If there is evidence of the desired protein being solubilized by this procedure, one can use a lower pH buffer such as 20 mM Tris–HCl, 1 mM EDTA pH 9.0.

Protocol 3

Washing of inclusion bodies with urea and detergents

Equipment and reagents

- Lab scale centrifuge: Sorvall RC5C or equivalent
- 1, 2, 3, and 4 M urea in 50 mM Tris–HCl, 1 mM EDTA, pH 8.0
- 0.1, 0.5, and 1.0% Triton X-100 in 50 mM Tris–HCl, 1 mM EDTA, pH 8.0
- Tekmar Ultra-turrax, Model TR5T
- 0.1, 0.5, and 1.0% octyl glucoside in 50 mM Tris–HCl, 1 mM EDTA, pH 8.0
- 0.1, 0.5, and 1.0% deoxycholate in 50 mM Tris–HCl, 1 mM EDTA, pH 8.0

Method

1 Add 20 ml of each wash buffer to individual aliquots of 1 g of inclusion body.

2 Thoroughly homogenize the inclusion bodies in each buffer using the Tekmar homogenizer.

3 Centrifuge the washed inclusion bodies at 10 000 g for 30 min at 4°C.

4 Collect the washed inclusion bodies.

5 Repeat steps 1–4 until the supernatant is clear or until other analytical methods ensure that no more of the contaminating proteins or nucleic acids are being

Protocol 3 continued

solubilized. Generally, three washes are sufficient to remove 90–98% of the contaminating materials.

6 Remove the wash additive by resuspending the washed inclusion bodies in water using the homogenizer with 20 ml of water for each gram of inclusion bodies.

7 Centrifuge as in step 3, collect the inclusion bodies, and repeat the water wash once more.

8 Ensure that the desirable protein from the inclusion bodies is not being solubilized by any of the wash solutions.

References

1. Williams, D. C., Van Frank, R. M., Muth, W. L., and Burnett, J. P. (1982). *Science*, **215**, 687.
2. Cousens, L., Shuster, J. R., Gallegos, C., Ku, L., Stempien, M. M., Urdea, M. S., *et al.* (1987). *Gene*, **61**, 265.
3. Parente, D., Deferra, F., Galli, G., and Gandi, G. (1991). *FEBS Lett.*, **77**, 243.
4. Thomas, C. P., Booth, T. F., and Roy, P. (1990). *J. Gen. Virol.*, **71**, 2073.
5. Modrich, P. and Zabel, D. (1976). *J. Biol. Chem.*, **251**, 5866.
6. Botterman, J. and Zabean, M. (1985). *Gene*, **7**, 229.
7. Schein, C. H. (1989). *Bio/Technology*, **7**, 1141.
8. Wilkinson, D. L. and Harrison, R. G. (1991). *Bio/Technology*, **9**, 443.
9. Chrunyk, B. A., Evans, J., Liliquist, J., Young, P., and Wetzel, R. (1993). *J. Biol. Chem.*, **268**, 18053.
10. Wetzel, R. (1994). *TIBTECH*, **12**, 193.
11. Speed, M. A., Wang, D. I. C., and King, J. (1995). *Protein Sci.*, **4**, 900.
12. Teschke, C. M. and King, J. (1995). *Biochemistry*, **34**, 6815.
13. Georgiou, G., Valax, P., Ostermeier, M., and Horowitz, P. M. (1994). *Protein Sci.*, **3**, 1953.
14. Wong, H. H., O'Neill, and Middleberg, A. P. (1996). *Bioseparations*, **6**, 185.
15. Forman, S. M., DeBernardez, R. S., Feldberg, R. S., and Swartz, R. W. (1990). *J. Membrane Sci.*, **48**, 263.
16. Meagher, M. M., Barlett, R. T., Rai, V. R., and Khan, F. R. (1994). *Biotechnol. Bioeng.*, **43**, 969.
17. Maachupalli-Reddy, J., Kelley, B. D., and De Bernardez Clark, E. (1997). *Biotechnol. Prog.*, **13**, 144.
18. Langley, K. E., Berg, T. G., Strickland, T. W., Fenton, D. M., Boone, T. C., and Wypych, J. (1987). *Eur. J. Biochem.*, **163**, 313.
19. Nambiar, K. P., Stackhouse, J., Presnell, S. R., and Benner, S. A. (1987). *Eur. J. Biochem.*, **163**, 67.
20. Kupper, D., Reuter, M., Mackeldanz, P., Meisel, A., Alves, J., Schroeder, C., *et al.* (1995). *Protein Expression Purification*, **6**, 1.
21. Boismenu, R., Semeniuk, E., and Murgita, R. A. (1997). *Protein Expression Purification*, **10**, 10.
22. Sugimoto, S., Yokoo, Y., and Hirano, T. (1991). *Agric. Biol. Chem.*, **55**, 1635.
23. Weigel, U., Meyer, M., and Sebald, W. (1989). *Eur. J. Biochem.*, **180**, 295.
24. Koths, K. E., Thomson, J. W., Kunitani, M., Wioson, K., and Hanisch, W. H. (1985). *EP* 0156373.

25. Smith, A. T., Santamu, N., Daily, S., Edwards, M., Bray, R. C., Thorneley, R. F., *et al.* (1990). *J. Biol. Chem.*, **265**, 13335.
26. Bowden, G. A., Paredes, A. M., and Georgiou, G. (1991). *Bio/Technology*, **9**, 725.
27. Valax, P. and Georgiou, G. (1993). *Biotechnol. Prog.*, **9**, 539.
28. Rinas, U. and Bailey, J. E. (1992). *Appl. Microbiol. Biotechnol.*, **37**, 609.
29. Hartley, D. L. and Kane, J. F. (1988). *Trends Biotechnol.*, **6**, 95.
30. Kane, J. F. and Hartley, D. L. (1991). In *Purification and analysis of recombinant proteins* (ed. R. Seetharam and S. K. Sharma), 121.
31. Schoner, R. G., Ellis, L. F., and Schoner, B. E. (1985). *Bio/Technology*, **3**, 151.
32. Allen, S. P., Polazzi, J. O., Gierse, J. K., and Easton, A. M. (1992). *J. Bacteriol.*, **174**, 6938.
33. Hellebust, H., Murby, M., Abrahmsen, L., Uhlen, M., and Enfors, S.-O. (1989). *Bio/Technology*, **7**, 165.
34. Babbitt, P. C., West, B. L., Buechter, D. D., Kuntz, I. D., and Kenyon, G. L. (1990). *Bio/Technology*, **8**, 1083.

Chapter 3
Purification for crystallography

S. P. Wood and A. R. Coker

University of Southampton, School of Biological Sciences, Biomedical Sciences Building, Bassett Crescent East, Southampton SO16 7PX, UK.

1 Introduction

In the early days of protein biochemistry, crystallization was often employed as a purification technique and crystallinity was considered an index of purity. Crystallographic studies were focused on a selection of proteins whose principle qualification was their abundance and ease of crystallization. The techniques of protein crystallography have now developed to such a degree that a complete three-dimensional structure analysis may not be a long task. Indeed it is often the production of suitable crystals that is rate limiting for the study of many proteins. High purity of the protein preparation used for crystallization is often one of the most important factors for growing diffraction quality crystals. Many crystallization trials have run into difficulties due to protein preparations which are not as homogeneous as first thought. Even for those proteins that can be purified from very impure mixtures by crystallization, crystal quality improves with each recrystallization as the purity increases. Crystal growing procedures and methods of purification have developed to cope with ever smaller quantities of protein derived from expression systems in bacteria, yeast, fungi, and cultured eukaryotic cells. This has allowed structure investigations of molecules that do not occur in nature or molecules or fragments whose natural abundance is very low. These expression systems often generate particular challenges in purification. The relationship of homogeneity and crystal growth is not well documented in the literature. Reports of successful crystal growth are numerous, but many of the problems encountered are not fully recorded. Failure to *consistently* grow crystals of high quality is a problem that afflicts many structure analyses. In view of the simplicity of crystal growing techniques and the relative complexity of many purification procedures, it is a reasonable supposition that the variability often derives from the latter. However this is not always the case and it is not an uncommon observation that apparently identical crystallizations made from the same batch of protein can yield crystals of variable quality or no crystals at all! A number of extensive texts are available that discuss crystal growth and the many techniques and tricks used to obtain diffraction quality protein crystals (1–5).

It is useful to consider the nature of protein crystals and how they are formed in order to examine the importance of homogeneity. The protein crystal is a rather open three-dimensional lattice, where each repeating motif is a single protein molecule or group of molecules. Much of the crystal volume is occupied by water molecules (30–90%) and only a small portion of the protein surface is involved in contacts with other protein molecules. Those water molecules close to the protein surface are well organized in hydrogen bonding networks with the protein and with each other, while further out, in the solvent channels of the crystal, their properties are more like those of bulk water. The types of interactions involved in contacts between protein molecules are very much like those believed to stabilize protein structure, namely, multiple weak forces involving hydrogen bonding, ion pairs, hydrophobic interactions, and occasionally metal co-ordination. Analysis of the amino acid composition of the surface regions of proteins involved in crystal contacts shows that these do not differ significantly on the average from the usual residue distributions on protein surfaces. This contrasts with the situation in protein oligomers where regions of atypical composition are involved in more extensive and stable contacts (6). For membrane proteins, the lattice also has protein–detergent and micelle-like detergent–detergent interactions. As a result all protein crystals are soft and extremely sensitive to environmental factors such as humidity. Classically, a crystal is defined by the almost faultless repetition of these protein–protein contacts involving perhaps 10^{16} molecules. However, recent atomic force microscopy images of protein crystals suggest that while there is more or less global order to the crystal, many local aberrations (dislocations, change of lattice) are visible (7) and these must account in part for the relatively weak diffraction and broad rocking width of diffraction spots of many protein crystals.

The basic principles of protein crystallization are the same in many respects as those which form the foundation for the more familiar techniques of crystallizing small molecules like salts or amino acids. It is necessary to achieve a slow approach to a low degree of supersaturation of protein in solution, that is, a condition in which the solvent holds more protein in solution than is possible at equilibrium. This metastable protein solution is usually achieved by a gradual change of precipitant or protein concentration, or by a change of pH or temperature. Naturally consideration must be given to what effect these changes may have on the protein structure/activity, and for membrane proteins one also needs to be aware of the effects such changes have on the state of the detergent. The final conditions of the crystallization should be such that the crystalline state is thermodynamically most stable. During searches for appropriate conditions, amorphous states are achievable from a sometimes distressingly wide variety of conditions, reflecting their probable random bonding network and kinetic accessibility. The crystalline state when achieved is usually of far lower solubility than the closest amorphous state, and sometimes crystal growth is sustained at the expense of redissolving of coexisting amorphous material. Techniques of dialysis, vapour diffusion, and slow cooling or warming are used to approach the required conditions. In such conditions of low solubility, appropriate molecular aggregates slowly reach a critical size, forming a nucleus from which crystal growth proceeds as a spontaneous process. The slow approach and low degree of supersaturation are important since the slow diffusion rates of macromolecules will limit the rates of presentation of appropriately oriented molecules to growth points and the number of such nuclei is related to super-saturation. The degree of supersaturation required to initiate homogenous nucleation (that is, nuclei generated from protein aggregates) is considered to be much greater than that required to sustain growth. This explains why it is often possible to produce showers of small crystals from excess nucleation, but much more difficult to produce a small number of larger crystals. Fortunately, at low degrees of supersaturation nucleation often appears to be heterogeneous, with growth initiating on particles of dust, fibre, or hair which in spite of the experi-menters efforts always seem to gain entry to the crystallization.

During protein purification, valuable information may be gained about the efficacy of different protein precipitants and about the stability of the protein. Limits of protein solubility with pH and estimates of isoelectric point can be determined. Changes in apparent molecular weight with conditions might pro-vide clues to specific aggregation processes necessary for crystal packing. Any factor that influences protein solubility might be manipulated to encourage a marginal dominance of the appropriate protein–protein interactions. Most pro-teins exhibit a solubility minimum at their pI where net charge is zero and repulsive forces are minimal. Many crystals grow close to these conditions. How-ever, proteins are very complex polyions with particular patterns of charge on their surface which give rise to multiple solubility minima. This is reflected in the growth of many protein crystals several pH units away from the pI.

Locating and optimizing the conditions required to produce good crystals can still be a difficult task in view of the large number of conditions that may need to be searched. If we start out with the assumption that there exists an optimum condition somewhere in the multi-dimensional parameter space defined by pH, ionic strength, counter ions, temperature, etc. and that there is no reason to suggest that these can be optimized individually, then a very large number of experiments is required to test all combinations. The statisticians approach to solving this 'needle in a hay stack' problem has dramatically changed the way we search for crystallization conditions (8–11). This involves setting up a rel-atively small number of experiments, an incomplete factorial, where random combinations of experimental conditions are selected that range over the entire parameter space. Results from this small subset of the large number of possible experiments should be sensitive to the influence of major factors and factor combinations through averaging the results of non-identical experiments that share particular features. The main experimental difficulty lies in assigning scores to experimental results, particularly at the early stages of an investigation when an array of qualitatively different precipitates may be found and quanti-fying one's intuition or experience of the potential for certain precipitates to lead on the right path is required. Once crystals have been observed then it is much easier to score the development of desirable characteristics. Furthermore, once a 'hit' has been located then it is possible to surround the hit with a full factorial experiment or some other local search method, testing possible combinations of factors to seek the direction of the optimum with respect to the variables.

Another search procedure suggested by Kim and co-workers (12) employs a 'sparse matrix' of 98 conditions that reflect the range of conditions that have proved successful in the past. This information is available in a crystallization database published on the world wide web (http://h178133.carb.nist.gov:4400/bmcd/bmcd.html) (13). The appropriate solutions are available commercially (Hampton Research) and the procedure has proved so successful that one would be unwise not to try this approach. Having set up the screen by vapour diffusion or microbatch (see below) then very careful study with a microscope is required to detect the few tiny microcrystals that may develop when the screen conditions approach an interesting area of the parameter space. Although on occasions the screen will take one close to an ideal and produce unmistakable crystals, more often the pH, ionic composition, protein concentration, etc. will be some way from an optimum. *Table 1* shows a simple experiment that might follow on from a sparse matrix hit to examine two levels off each of four variables. When moving away from a hit it is important to study carefully how the commercial solutions have been prepared in order to make reasonable modifications. In particular it is worth noting that the pH quoted for a particular condition is based on the starting stock buffer and not the pH of the final cocktail of reagents. Alternative buffers for achieving the same pH are worth including in optimization as their concentrations are often a couple of orders of magnitude higher than the protein and weak interactions may well be important. It is important to be mindful of some basic inorganic chemistry when setting up screens as both phosphate ions and magnesium ions feature in a number of the screen solutions and if either exist in the sample then very fine

Table 1 The layout of a crystallization tray suitable for screening four variables at two different levels each[a]

Precipitant					Protein concentration
		pH			
	Low		High		
Low	A2	A3	A4	A5	**Low**
	B2	B3	B4	B5	**High**
High	C2	C3	C4	C5	**Low**
	D2	D3	D4	D5	**High**
	Low	**High**	**Low**	**High**	
		Additive			

[a] Each well is labelled A–D vertically and 2–5 horizontally, corresponding to the central block of wells on a Linbro plate, and the variables with their set levels are labelled around the edge. If we consider the four wells in the top left-hand corner of the tray we see that they all have the same pH and precipitant concentration but differ in the amount of additive and the protein concentration. By comparing these four wells we can clearly see the effect of the additive and the protein concentration. By comparing the wells in row A we can clearly see if any effect of the additive is dependent on pH, and so on.

magnesium phosphate crystals will grow and lead to premature celebrations of success. The success of this screening approach to crystallization has spawned many local variants on the same theme and a range of special screens for membrane proteins, nucleic acids, DNA binding proteins. If you still have no crystals, then it is time to look more carefully at what you know about the protein and its properties.

2 Methods for protein crystallization

2.1 Batch

The traditional approach to crystallization is the 'batch' method which involves adding precipitant or changing pH until the protein comes to the limit of its solubility. With luck, a few nuclei will be generated and if the solution is left in a quiet corner large crystals will grow. The major limitation of batch experiments is that precipitant conditions remain static. The only change in the time course of the experiment is a decrease in the protein concentration. This can aid the crystallization by moving the system out of the nucleation zone and preventing over nucleation, but each experiment only surveys one set of precipitant conditions so the crystallization conditions need to be known rather precisely before setting up the experiment. The traditional way of overcoming this limitation was to slowly decrease the solubility of the protein by allowing a

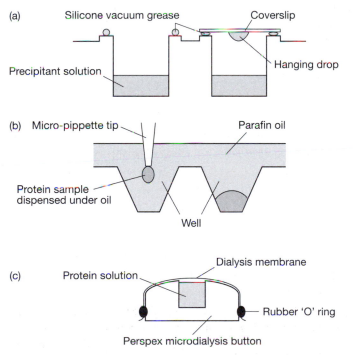

Figure 1 Schematic diagrams of (A) a hanging drop experiment, (B) a microbatch experiment, and (C) a micro-dialysis button.

warmed batch experiment, insulated inside a Dewar, to slowly cool in a cold room.

The large volumes (100 μl and up) used in the traditional batch method make it suitable for growing large crystals but make it impractical for screening crystallization conditions. In a more recent interpretation, the 'microbatch' method (14), a droplet of protein solution is immersed in paraffin oil (see *Protocol 1, Figure 1*). This prevents evaporation of the solution so very small volumes (down to 1 μl) can be used making the method ideal for initial screening. A further refinement of the technique uses a mixture of paraffin oil and silicone oil. The silicone oil allows slow diffusion of water, and so the concentration of protein and precipitant in the drop slowly increase as water evaporates. In this way, a range of protein and precipitant concentrations are automatically surveyed in each droplet. The disadvantage of this is that the drops will dry out in a matter of weeks, unless stored between 4 and 10 °C, so they must be examined regularly for crystals.

Protocol 1

Microbatch

Equipment and reagents

- Terasaki plate (Hampton Research): these plates contain 72 cone shaped wells that can hold up to 10 μl
- Oil–liquid colourless light paraffin (BDH)
- Precipitant solutions
- Protein solution, centrifuged to remove particulates

Method

1 Fill the Teraski plate with enough oil to cover all of the wells.

2 Add 1–5 μl of each precipitant solution to each well. The drop of precipitant can be dispensed just under the surface of the oil and will sink to bottom of the well.

3 Add 1–5 μl of the protein solution to each well in the same way. The protein solution and precipitant can be allowed to mix by diffusion.

4 Store plate in a quiet place, preferably with controlled temperature.

5 Inspect plate after one and two weeks.

6 Inspect plate each month for the next six months. The aqueous solution in the drops will slowly evaporate through the paraffin oil increasing the concentration of protein in the drop.

7 Crystals can be harvested by draining excess oil from the plate so that only the wells are filled with oil. Then wells containing crystals can be filled with 15 μl of the appropriate precipitant solution which will displace the oil. The crystals can now be mounted as usual.

Modifications

1 Some people feel it is more convenient to add the protein solution and precipitant prior to adding the oil. The oil can then be on top of the crystallization drop with a micropipette. It should be noted that thinner layers of oil will increase the evaporation rate from the drop.

2 The evaporation rate from the drop can be regulated by mixing silicone oil with the paraffin oil (15). Increasing amounts of silicone oil (e.g. Dow Corning silicone fluid 200/litre cS—from BDH) allows the drop to evaporate more quickly and in this way vapour diffusion can be mimicked.

2.2 Vapour diffusion

Vapour diffusion has been the most widely successful method used for producing diffraction quality crystals to date. In this method a drop of protein solution mixed with precipitant, is slowly dehydrated, in a sealed well, by equilibration with a reservoir at higher precipitant concentration. The protein drop can either be suspended from the coverslip used to seal the reservoir (a hanging drop, *Figures 1* and *2*) or sat on some sort of support above the reservoir (a sitting drop). Usually the drop is made by mixing equal volumes of protein solution and reservoir solution. Vapour diffusion has the benefit of automatically screening a range of precipitant concentrations but, unlike a dehydrating microbatch experiment, it has a defined end-point. In theory, this should allow one to optimize a crystallization so that the precipitant concentration slowly reaches the nucleation region of the solubility phase diagram, and thereafter increases just fast enough to compensate for loss protein to the crystals. This requires not only the optimization of the end-point of the experiment but also the rate of equilibration. The simplest way to vary the rate of equilibration is to vary the surface to volume ratio (i.e. the size) of the drop. In general larger drops will equilibrate more slowly than smaller drops, for example, a 32 μl hanging drop will take more than twice as long to equilibrate as an 8 μl drop (16). Another easy way to vary the time course of the experiment is to pipette a 200 μl layer of paraffin/silicone oil on top of the well solution (17). This will slow the equilibration to a degree dependent on the ratio of paraffin to silicone oil. The precipitant used plays a major role in kinetics of the drop equilibrium. Wells with salt precipitants such as ammonium sulfate can reach equilibrium in one to two days whereas wells containing PEG precipitants alone can take more than a month (16). Consequently crystals that grow after a few days in a PEG only crystallizations could be considered analogous to microbatch experiments. The addition of modest amounts (e.g. 200 mM) of NaCl will reduce the equilibration time of a PEG crystallization to around 10 days (18), and may be beneficial in such cases. However, a general (but not exclusive) observation is that long equilibration times give better quality crystals.

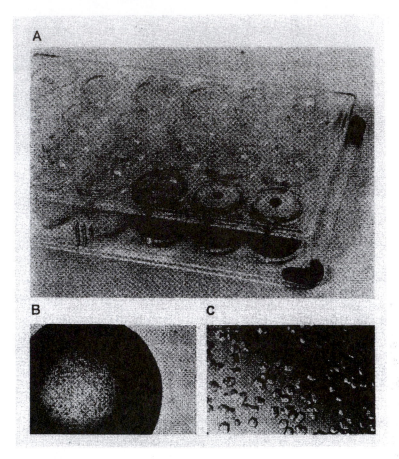

Figure 2 (A) The experimental set-up for a hanging drop vapour diffusion crystallization. A Linbro plate provides 24 wells for 0.5–1 ml of precipitant reservoir. Protein drops hang from the coverslip which is sealed to the well rim with silicone vacuum grease. A small spacer should always be employed (here on the corner of the plate) to separate the lid from the coverslips. Coloured solutions have been used for the well and drop contents to aid inspection. (B, C) Small crystals (< 100 μm) of the aspartic proteinase of avian myeloblastosis virus produced in such an apparatus at low and higher magnification to show what one might expect to see at a successful initial screen point (courtesy of J. Cooper and P. Stropp).

Protocol 2

Vapour diffusion in hanging drops

Equipment and reagents

- Linbro tissue culture tray (Flow Laboratories)

- Precipitant solutions

- Silicone vacuum grease, in a 20 ml syringe

- Siliconized coverslips, 22 mm, square (Hampton Research)

- Protein solution, centrifuged to remove particulates

Protocol 2 continued

Method

1   Pipe silicone grease around the rim of each well of the Linbro tray. This will provide a seal with the coverslip when closing the well.

2   Dispense and mix 1 ml of precipitant solution into each well.

3   Place 2–8 μl of protein solution in the centre of a coverslip.

4   Carefully layer 2–8 μl of precipitant solution from a well on top of the protein solution drop.

5   Invert the coverslip over the same well and press into the silicone grease to create a seal.

6   Store in a quiet place, preferably with temperature control.

Modifications

1   The rate of equilibration may be controlled in several ways:

   (a) Varying the size of the drop, larger drops will equilibrate more slowly.

   (b) Adding a layer of oil to the top of the reservoir will slow the equilibration, a suitable oil would be an equal mixture of paraffin and silicone oils (14).

2   Sitting drops may be used rather than hanging drops. The drop sits on a platform rather than hanging from a coverslip. Sitting drops slow down the equilibration, allow larger drop sizes, and prevent large delicate crystals being distorted by the curvature of the drop surface. A convenient method for setting up sitting drops is to use a Micro-Bridge (Hampton Research) which sits neatly inside the wells of a Linbro tray.

2.3 Dialysis

Dialysis procedures have also been quite successful. Again a range of precipitant conditions are automatically screened but unlike the previous two methods the protein concentration remains constant. Semi-permeable membranes of the types commonly employed in preparative procedures are used to contain the protein solution in thick-walled capillary tubes or Perspex micro-dialysis buttons (see *Figure 1* for illustration). The membrane is secured with an 'O' ring or gaiter cut from a suitable diameter of silicone rubber tubing. The device is then suspended or immersed in a dialysable precipitant, a low ionic strength buffer or water. The exterior conditions can be varied in a stepwise manner, with an equilibration interval of a few days between steps. In capillary tubes, gradients of precipitant strength in the column of protein solution may lead to crystal growth at a particular height and aid choice of the ideal precipitant strength. In general, dialysis methods are not economical on protein. Perspex buttons with an internal volume of 5 μl are available, but more than this is usually required to successfully fill a button without trapping air bubbles under the membrane. However, dialysis methods do allow more flexibility as amorphous precipitates in failed attempts to produce crystals can be redissolved with appropriate solution placed outside of the chamber, and another condition investigated with the same sample.

3 Influence of heterogeneity on crystallization

3.1 Effects of impurities

Lack of purity in the protein sample may lead to:

(a) Complete failure to crystallize.

(b) The production of only small crystals.

(c) Large crystals which do not scatter X-rays well.

If the protein has not previously been crystallized, then the degree of purity becomes an additional variable to be screened. Contaminating species which are able to satisfy only a portion of the bonding interactions necessary to propagate the lattice may be excluded during crystallization. Any variation in the test molecule in these contact areas will at best reduce the pool of material in the crystallization that can be incorporated into the crystal. In a worst case, where perhaps only a fraction of the crystallization 'valencies' of the protein are damaged, the defective molecules might be built into the lattice and lead to a halt in further growth or disturbances in the lattice. Accumulated defects are believed to be a major factor in limiting crystal size. One implication of this argument is that microheterogeneity in the protein of interest is probably more troublesome than contamination from some totally unrelated molecule. However, recrystallization is useful in some circumstances for purification where the demands of packing in the crystal lattice are utilized to select out only those molecules with the appropriate structure. For instance, crystallization of insulin was of major importance in producing higher quality preparations for the treatment of diabetes mellitus (19). Nevertheless, we now know that repeated crystallization of insulin fails to remove a residual contamination of about 5% comprising of proinsulin, various intermediates of its conversion reaction, and chemically modified forms from the extraction process. These components can be removed by gel filtration, ion exchange, and reversed-phase chromatography. This microheterogeneity presents a more significant purification problem, since variation in properties may be rather slight. It is likely to be encountered to some degree with most protein preparations, reflecting common features of incomplete biosynthetic processing, natural turnover, and abuse during extraction. Even commercial sources of hen egg white lysozyme that easily produce high quality crystals have recently been carefully characterized and shown to contain small quantities of ovalbumin and lysozyme dimers (20). X-ray topography measurements suggest that such impurities in crystals cause considerable strain in the crystal lattice which reduces the quality of diffraction (21). An interesting attempt to simulate the effects of microheterogeneity has been provided by experiments in which hen egg white lysozyme was used as a contaminant in crystallizations of turkey egg white lysozyme (22). The two proteins have a 95% sequence identity with all the differences located on the surface of the molecules. The effect of this artificially induced microheterogeneity was to alter the crystal morphology by inhibiting growth along one face of the crystal. In general such an effect could result in needle or thin plate like crystals unsuitable for diffraction experiments.

3.2 Origins of microheterogeneity

Microheterogeneity might derive from deamidation of asparagines or glutamines, cysteine oxidation, proteolysis, sequence polymorphism, variability in post-transitional modifications like serine phosphorylation or N-methylation or acetylation, and sialic acid capping of oligosaccharide chains. Some examples are described.

3.2.1 Deamidation

One of the six side chain amides of bovine insulin deamidates much faster than others at extremes of pH. The product is fully active and crystallizes isomorphously with undegraded insulin (23). However, the crystal morphology bears little resemblance to the typical rhombohedral form of intact insulin due to a disturbance of the relative growth rates of crystal faces. If an insulin preparation is contaminated heavily with the deamidated form, then curious 'snowflake' polycrystalline aggregates can form in crystallization.

3.2.2 Oxidation

The reactivity of free thiols varies widely with solvent exposure. Frequently an enzyme active site geometry will enhance the reactivity of such groups which, unless protected, will then be very effective scavengers of trace metals in preparation buffers or readily become oxidized, leading to inactivation and possibly denaturation. The ability of crystals of thymidylate synthase from *Lactobacillus casei* to diffract X-rays has been related to a thiol oxidation state (24). Dimers of reduced or reduced and oxidized forms of the enzyme appear to produce good crystals, while complete oxidation or a random distribution of redox states obtained on exposing preformed crystals to air leads to disorder. Successful control of thiol oxidation was shown to be crucial for reproducible production of crystals of β-hydroxybutyrate dehydrogenase (25).

An interesting example of a more complex oxidation problem is provided by studies of the enzyme porphobilinogen deaminase from *E. coli* (26). The enzyme crystallizes readily at pH 5.5 by PEG precipitation. However, the enzyme's covalently bound dipyrromethane cofactor readily oxidizes at this pH, giving an inactive protein with a yellow colour. The partially oxidized protein produces large crystals that provide X-ray data to 1.7 Å resolution. It is more difficult to produce large single crystals with the unoxidized form of the protein, although useable crystals were grown in the presence of a large excess of DTT. The explanation for this is that the oxidized cofactor has a rigid planar structure that seems to have an organizing influence on the surrounding protein domains, whereas the unoxidized cofactor is more flexible. The sting in the tail is that the oxidized protein is very sensitive to light. It seems that on absorption of light the oxidized cofactor may undergo an isomerization reaction, analogous to that in the visual pigment retinal. This disrupts its interactions with the surrounding protein, apparently destabilizing the entire structure. Consequently the crystals of the oxidized form of the protein can only be grown in the dark. Once these crystals have been grown, exposure to normal laboratory light for a few days is sufficient to completely disorder the crystals and give them a malleable texture.

3.2.3 Phosphorylation

The activity of glycogen phosphorylase is partly controlled by the phosphorylation state of a single serine residue near to the N-terminus. The phosphorylated a-form is more active. Fortunately, in resting muscle the a-form predominates, and the enzyme which is responsible for conversion to the b-form can be quickly removed from extracts. Alternatively, the b-form may be accumulated. Thus at an early stage the potential difficulties of separating chemically rather similar species are minimized. Glucose-inhibited a-form and IMP-activated b-form crystals have been studied, and show conformational differences, particularly in the region of the phosphorylation site (27–29).

3.2.4 Proteolytic processing

Preparations of mouse submaxillary gland nerve growth factor show variable chain length following proteolytic processing from a large precursor (30). Eight residues from the N-terminus and one from the C-terminus can be missing. The situation is further complicated since the protein is isolated as a stable dimer with many possible chain length combinations. The isoelectric points of the variants are very close, but the X-ray structure is not informative at the N-terminus (31).

3.2.5 Isoenzymes

The fungus, *Rizopus chinensis*, produces two forms of an extracellular aspartic proteinase which have pI values of 5.6 and 6.0. When co-purified, reasonable crystals can be grown. Following separation by preparative isoelectric focusing, the form with pI 5.6 produces larger crystals which diffract X-rays to a higher resolution and are less sensitive to radiation damage (32). The form with pI 6 has not been crystallized.

3.2.6 Carbohydrate constituents

Human leukocyte elastase shows five bands on gel electrophoresis which have been attributed to carbohydrate heterogeneity, but the protein crystallizes well (33). Neuraminidase may be used to remove terminal sialic acid residues from oligosaccharide chains. These charged residues are probably responsible for the multiple bands seen on gels. Numerous other glycosylated serum proteins have also been successfully crystallized. In spite of this, the carbohydrate attached to proteins is often viewed with suspicion by those trying to grow crystals. This is partly due to its tendency to be heterogeneous, but also perhaps to our poor understanding of the role of carbohydrate. Sometimes the chains are not visible in electron density maps, but in other cases such as immunoglobulins (34), influenza virus coat haemagglutinin (35), and neuraminidase (36), some chains are clearly defined, particularly when involved in protein–protein contact. Both viral proteins are heavily glycosylated (20% by weight). The sugar chains are free of sialic acid, due to the neuraminidase activity. A range of enzymes is now available for the removal of carbohydrate chains (Oxford Glycosystems) but the cost of this process can be considerable and removal of the cleavage products difficult. Sugar cleaving enzymes have been expressed as fusion proteins with glutathione-S-transferase enabling their rapid removal from the deglycosylated protein by affinity chromatography on a glutathione column (37). In the case of human chorionic gonadotrophin (HCG) crystallization and structure analysis was enabled by partial carbohydrate removal with HF (38–40). Proteins expressed in micro-organisms following genetic manipulation provide a useful source of carbohydrate-free proteins, as complete removal from existing glycoproteins is not easy. Interferon-γ provides a recent example (41). Similarly membrane anchor regions that lead to solubility problems can be deleted from the DNA of the protein to be expressed.

3.3 Detection of microheterogeneity

Microheterogeneity involving a change of a charge in the contact region between molecules in the crystal is likely to be deleterious. Analytical isoelectric focusing on narrow and broad pH ranges together with polyacrylamide gel electrophoresis at pH values above and below the pI are standard procedures capable of detecting most potential problems. Internal chain cleavages may be detected by SDS electrophoresis in reducing conditions. Protein preparations should always be analysed by these methods before crystallization, as high performance chromatographic supports for ion exchange and chromatofocusing can usually provide the resolution necessary to prepare materials of uniform charge. The criteria of purity in all chromatographic separations should be based on the production of symmetrical elution peaks. In ion exchange chromatography this is achieved with gentle gradients of the eluting counter ion and in size exclusion chromatography by ensuring that the correct fractionation range is employed. Complex mixtures can appear rather pure if eluted with steep salt gradients or eluting in the void volume of a size exclusion column. Silver staining procedures for analytical electrophoresis methods are in many cases sufficiently sensitive to display purity limits governed by protein stability and handling techniques. A single silver stained band in a heavily loaded gel is rarely achievable. A very powerful technique that is becoming more widely available for protein characterization is electrospray mass spectrometry. With as little as 50 μg of thoroughly desalted protein it is possible to obtain very precise estimates of the protein molecular mass ($\pm$ 0.01%) that enable rather specific

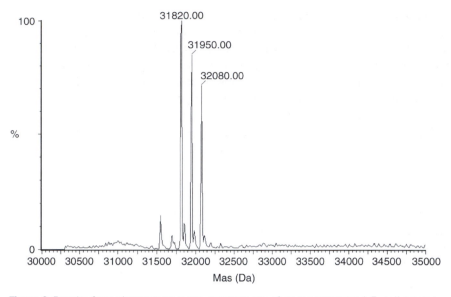

Figure 3 Results from electrospray mass spectrometry of an overexpressed *E. coli* enzyme showing heterogeneity due to incomplete methionine removal at the N-terminus, the mature enzyme is 31 820 Da. The preparation appears homogeneous by ion exchange and gel filtration chromatography and shows a single band on native and SDS gel electrophoresis.

identification of problems with the covalent structure of the protein that are potentially damaging to crystallization, such as N-terminal methionine removal from proteins expressed in *E. coli* (42). *Figure 3* shows an example of the power of electrospray mass spectrometry. Here an overexpressed *E. coli* hydrolase has undergone incomplete N-terminal processing leaving none, one, or two methionines at the N-terminus.

Where variability in sequence involves no change of charge, reverse-phase chromatography has proved extremely powerful for smaller proteins. For instance, porcine pituitary gland neurophysin I preparations contain a species in which a single C-terminal leucine deletion occurs, and this was first detected by reverse-phase chromatography (43). Similarly, bovine pancreatic proinsulin exhibits a single Leu–Pro polymorphism which is resolvable (44). Of course, at the beginning of crystallization trials one cannot predict how damaging a particular contaminant might be. The various types of heterogeneity outlined above may fortuitously fall outside contact regions between molecules and be tolerated in the lattice. Neurophysin and proinsulin both crystallize fairly well (45, 46) without the separation of the sequence variants, while the two insulins of the rat do not, since they crystallize in the same conditions to produce a rhombohedral and a cubic packing arrangement (47). Mutations may be introduced to recombinant proteins to improve order in the crystal lattice (42) and on occasions accidental mutations from PCR amplification of DNA have proved of great assistance in improving crystals of the recombinant protein (48).

Heterogeneity may, on the other hand, accompany partial unfolding of the protein or promote distinct conformational isomers in more flexible proteins which will not be so readily detected. Denaturation might only be observed as a change of solubility, a slight alteration in apparent molecular size during gel filtration chromatography, or an increased susceptibility to proteolysis. This kind of damage quite often leads to severe non-specific aggregation problems and this can be detected very effectively by dynamic light scattering (49). Large particles scatter laser light in the visible wavelength region strongly, enabling detection of very small levels of contamination. Such aggregates are considered to be severely deleterious to crystal growth. The light scattering detectors enable searching for solution conditions to inhibit aggregation. Glycerol is often an effective additive to control aggregation (50, 51). Freeze-drying was found to be detrimental to the crystallization of carbonic anhydrase, presumably through partial denaturation. Many enzymes need to be protected from the denaturing effects of ice crystals by storage at low temperature in glycerol.

A particularly devious problem can occur when expression is being driven with high efficiency but the cellular folding machinery cannot cope. This can lead to inclusion body formation but also to soluble partially folded protein. Removal of such material is a particularly important application of affinity chromatography. High affinity ligand binding sites on proteins are very sensitive to overall structural changes. On the other hand, affinity chromatography will be insensitive to a host of other features of microheterogeneity since it targets only a small region of the protein that will often be in a cleft or pocket. Affinity tags such as the histidine tails used so effectively for fishing out recombinant product from the soup of an *E. coli* lysate by Ni-chelate chromatography are not sure to provide ideal material for crystallization without further purification. The histidine tails themselves are capable of adding a substantial charge to the protein that will vary with pH and affect solubility. Nevertheless, many proteins have been crystallized with the tag in position. Most commercially available constructs include a proteolysis site to enable tag removal. On occasions, the subtleties of a particular state of purity may not be evident (for example in the exact lipid–detergent composition associated with a solubilized membrane protein complex), and in such cases rigorous reproduction of purification conditions is essential, particularly once crystals have been grown. In fact, rather few membrane proteins have been successfully crystallized. The methods used are very similar to those outlined in Section 2, but detergent is an obligatory constituent. Other small amphiphilic additives are also used (52–55).

3.4 Additives and proteolysis

In contrast to the previous sections, there are many examples in which non-peptide species other than precipitants are important in crystallization. Cofactors, allosteric effectors, and metal ions required for enzyme activity are perhaps the easiest to identify as specific activity is often monitored to follow purification. Furthermore, such compounds can be protective and are usefully included in preparation buffers. Enzyme substrates, products, and carrier protein ligands have also been useful. Malic enzyme from rat liver, for instance, requires NADP for crystallization (56). Enzyme inhibitors have been employed in the crystallization of *E. coli* dihydrofolate reductase (57) and mouse submaxillary gland renin (58). It often seems to be the case that proteins with extensive clefts or pockets that bind cofactors or substrates produce much better crystals when the ligand is bound. Presumably ligands stabilize the protein structure. The rate of crystal growth of serum amyloid protein can be controlled by the inclusion of a modified monosaccharide similar to that employed as an immobilized ligand on affinity chromatography supports during purification (59). Some caution is necessary, as binding of allosteric effectors to phosphofructokinase and aspartate carbamoyl transferase leads to substantial conformation change and subunit rearrangements (60, 61). Many enzyme crystals disintegrate in the presence of substrates and modifiers, indicating that the protein conformation is no longer consistent with the existing set of intermolecular contacts. Substantial changes in crystal form are found for various apo- and ternary complex forms of lactate dehydrogenase (62). Glyceraldehyde-3-phosphate dehydrogenase can be prepared with from 0 to 4 molecules of NAD bound (63). The crystallization of concanavalin A is inhibited by its ligand *N*-acetylglucosamine. It is clear that partial occupation of a protein preparation with such compounds could be a serious source of microheterogeneity. However, once under control, their binding provides the opportunity to investigate the relation of structure and mechanism.

In other circumstances, purification may unwittingly remove some essential component for crystallization. Chelating agents or dissociating conditions are often used to aid in combating proteolysis in crude tissue extracts, and these will remove metals required for crystal growth. The requirement for zinc ions in the crystallization of insulin is a well known example. Copper ions were found eventually to be essential for the crystallization of oxidized thioredoxin, although they are not known to be involved with the activity of the molecule (64). Cobalt ions were necessary to crystallize chloramphenicol acetyltransferase (65). In these examples, the metal ions seem to be important only in providing a link between molecules in the crystal lattice.

There are many examples (2) where controlled proteolysis leads to a species that is particularly suitable for crystallization (for example immunoglobulins, canavalin, elongation factor TU, ribonuclease S, cytochrome b5). In some cases, it seems that proteolysis is removing flexible or protruding portions of the protein to provide more compact structures which are easier to pack together. In other cases, linking strands of polypeptide can be cleaved to release intact domains. The ionophore domain of colicin A (66) and the Klenow fragment of DNA polymerase (67) are produced by thermolysin and subtilisin digestion for crystallization. Influenza virus haemagglutinin and neuraminidase provide interesting examples. Both native proteins are attached to the viral envelope membrane by a hydrophobic domain which is cleaved during preparation by bromelain or pronase! Their heavy glycosylation probably protects against more extensive proteolysis.

4 Concluding remarks

So far we have noted how microheterogeneity can have deleterious effects for the crystal lattice and it is not hard to envisage how the solubility properties of the protein might be blurred by such variability, leading to difficulties in defining crystallization conditions. For instance, traces of a contaminant or modified form of the protein of interest might be precipitated during a vapour diffusion equilibration before crystallization starts, producing an excess of nucleation sites. If crystals are obtained in spite of these factors, further problems may be waiting. Crystal morphology may be variable due to minor packing changes and resultant changes in the rate of development of crystal faces. The efficiency of heavy metal binding to the protein which is necessary for structure analysis might be impaired if, for instance, proteolysis or chemical modification has effectively removed important residues from a significant proportion of the protein molecules. Such degradation might also lead to poor definition in this part of the final structure. Correlations of structure and function may be led astray if chemical differences from the native form are not fully appreciated.

Crystal growth is not well understood, and almost entirely empirical rules are followed to grow crystals. Criteria of purity for crystallization should in practice be no more or less stringent than those tolerated for any other means of characterization of a pure protein. In as much as some methods of study need not necessarily be harmed by some lack of purity, the same may often be true of crystallization. However, the basis of the technique of X-ray analysis presumes identity of the molecules under study, and towards the end of a structure analysis the extent of any compromise at an early stage may be magnified to cause serious difficulties in interpretation.

Wherever rather special and complex recipes are required for crystal growth it is usually hard to outline a logical path to the eventual conditions or to see how systematic screening could cope alone. Rather, screening with a strong bias from prior knowledge of the properties of the protein gained during isolation seems the most powerful approach.

References

1. Blundell, T. L. and Johnson, L. (1976). *Protein crystallography*. Academic Press, New York.
2. McPherson, A. (1982). *Preparation and analysis of protein crystals*. John Wiley, New York.
3. Durcruix, A. and Giege, R. (ed.) (1992). *Crystallization of nucleic acids and proteins: a practical approach*. Oxford University Press.
4. Carter, Jr., C. W. and Sweet, R. M. (ed.) (1997). *Methods in enzymology*. Vol. 276. Academic Press.
5. McPherson, A. (1998). *Crystallisation of biological macromolecules*. Krieger.
6. Carngo, O. and Argos, P. (1997). *Protein Sci.*, **6** (10), 2261.
7. McPherson, A., Malkin, A. J., and Kuznetsov, Y. G. (1995). *Structure*, **3** (8), 759.
8. Carter, Jr., C. W. and Carter, C. W. (1979). *J. Biol. Chem.*, **254** (23), 12219.
9. Kingston, R. L., Barker, H. M., and Baker, E. N. (1994). *Acta Crystallogr.*, **D50**, 429.
10. Carter, Jr., C. W. and Yin, Y. (1994). *Acta Crystallogr.*, **D50**, 572.

11. Carter, Jr., C. W. (1996). *Acta Crystallogr.*, **D52**, 647.

12. Jancarik, J. and Kim, S. H. (1991). *J. Appl. Crystallogr.*, **24**, 409.

13. Gilliland, G. L., Tung, M., Blakeslee, D. M., and Lander, J. (1994). *Acta Crystallogr.*, **D50**, 408.

14. Cheyen, N. E. (1992). *J. Cryst. Growth*, **122**, 176.

15. D'Arcy, A., Elmore, C., Stihle, M., and Johnston, J. E. (1996). *J. Cryst. Growth*, **168**, 175.

16. Mikol, V., Rodeau, J., and Giege, R. (1990). *Anal. Biochem.*, **186**, 332.

17. Cheyen, N. E. (1997). *J. Appl. Crystallogr.*, **30**, 198.

18. Luft, J. R. (1995). *Acta Crystallogr.*, **30**, 198.

19. Blundell, T. L., Dodson, G. G., Hodgkin, D. C., and Mercola, D. A. (1972). *Adv. Protein Chem.*, **26**, 279.

20. Thomas, B. R., Vekilov, P. G., and Rosenberger, F. (1996). *Acta Crystallogr.*, **D52**, 776.

21. Vekilov, P. G., Monaco, L. A., Thomas, B. R., Stojanoff, V., and Rosenberger, F. (1996). *Acta Crystallogr.*, **D52**, 785.

22. Hirschler, J. and Fontecilla-Camps, J. C. (1996). *Acta Crystallogr.*, **D52**, 806.

23. Bedarkar, S. (1982). Ph.D. thesis, University of London.

24. Tykarska, E., Lebioda, L., Bradshaw, T. P., and Dunlap, R. B. (1986). *J. Mol. Biol.*, **191**, 147.

25. Drenth, J. (1988). *J. Cryst. Growth*, **90**, 368.

26. Jordan, P. M., Warren, M. J., Mgbeje, B. I. A., Wood, S. P., Cooper, J. B., Louie, G. V., *et al.* (1992). *J. Mol. Biol.*, **224**, 269.

27. Cori, G. T., Illingworth, B., and Keller, P. J. (1955). In *Methods in enzymology* (ed. S. P. Colowick and N. Kaplan), Vol. 1, p. 200. Academic Press, New York.

28. Fletterick, R. J., Sprang, S., and Madsen, N. B. (1979). *Can. J. Biochem.*, **31**, 339.

29. Weber, L. T., Johnson, L. N., Wilson, K. S., Yeates, D. G. R., Wild, D. L., and Jenkins, J. (1978). *Nature*, **274**, 433.

30. Server, A. C. and Shooter, E. M. (1977). *Adv. Protein Chem.*, **31**, 339.

31. McDonald, N. Q., Lapatoo, R., Murrayrust, J., Gunning, J., Woldawer, A., and Blundell, T. L. (1991). *Nature*, **354** (6352), 411.

32. Bott, R. R., Navia, M. A., and Smith, J. L. (1982). *J. Biol. Chem.*, **257**, 9883.

33. Bode, W., Wei, A.-Z., Huber, R., Meyer, E., Travis, J., and Neumann, S. (1986). *EMBO J.*, **5**, 2453.

34. Deisenhofer, J., Colman, P. M., Epp, O., and Huber, R. (1976). *Hope-Seyler's Z. Physiol. Chem.*, **357**, 1421.

35. Wilson, I. A., Skehel, J. J., and Wiley, D. C. (1981). *Nature*, **289**, 366.

36. Varghese, J. N., Laver, W. G., and Colman, P. M. (1983). *Nature*, **303**, 366.

37. Grueninger-Leitch, F., D'Arcy, A., D'Arcy, B., and Chene, C. (1996). *Protein Sci.*, **5**, 2617.

38. Lapthorn, A. J., Harris, D. C., Littlejon, A., Lustbader, J. W., Canfield, R. E., Machin, K. J., *et al.* (1994). *Nature*, **369**, 455.

39. Lustbader, J. W., Wu, H., Birken, S., Pollak, S., Kolks, M. A. G., Pound, A. M., *et al.* (1995). *Endocrinology*, **136**, 640.

40. Harris, D. C., Machin, K. J., Evin, G. M., Morgan, F. J., and Isaacs, N. W. (1989). *J. Biol. Chem.*, **264**, 6705.

41. Vijay-Kumar, S., Senadhi, S. E., Ealick, S. E., Nagabhushan, J. L., Trotta, P., Kosecki, R., *et al.* (1987). *J. Biol. Chem.*, **262**, 4804.

42. Oubridge, C., Ito, N., Teo, C.-H., Fearnley, I., and Nagai, K. (1995). *J. Mol. Biol.*, **249**, 409.

43. Schwandt, P. and Richter, W. O. (1980). *Biochem. Biophys. Acta*, **626**, 376.

44. Frank, B. H., Pekar, A. H., Pettee, J. M., Schirmer, E. M., Johnson, M. G., and Chance, R. (1984). *Int. J. Peptide Protein Res.*, **23**, 506.

45. Pitts, J. E., Wood, S. P., Hearn, L., Tickle, I. J., Wu, C. W., Blundell, T. L., *et al.* (1980). *FEBS Lett.*, **121**, 1.

46. Blundell, T. L., Pitts, J. E., and Wood, S. P. (1982). *Crit. Rev. Biochem.*, **13**, 141.
47. Wood, S. P., Tickle, I. J., Bludell, T. L., Wollmer, A., and Steiner, D. F. (1978). *Arch. Biochem. Biophys.*, **186**, 175.
48. Braig, K., Otwinowski, Z., Hegde, R., Biosvert, D. C., Joachimiak, A., Horwich, A. L., *et al.* (1994). *Nature*, **371**, 578.
49. Ferrè-D'Amarè, A. and Burley, S. K. (1994). *Structure*, **2** (5), 357.
50. Sousa, R. (1995). *Acta Crystallogr. D*, **51**, 271.
51. Gekko, K. and Timasheff, S. N. (1981). *Biochemistry*, **20**, 4677.
52. Michel, H. (1982). *J. Mol. Biol.*, **158**, 567.
53. Garavito, R. M., Jenkins, J., Jansonius, J. N., Karlsson, R., and Rosenbusch, J. P. (1983). *J. Mol. Biol.*, **164**, 313.
54. Michel, H. and Oesterhelt, D. (1980). *Proc. Natl. Acad. Sci. USA*, **77**, 1283.
55. Gros, P., Groendjik, H., Drenth, J., and Hol, W. G. J. (1988). *J. Cryst. Growth*, **90**, 193.
56. Baker, P. J., Thomas, D. H., Howard Barton, C., Rice, D. H., Howard Barton, C., Rice, D. W., *et al.* (1987). *J. Mol. Biol.*, **193**, 233.
57. Mathews, D. A., Alden, R. A., Bolin, J. T., Freer, S. T., Hamlin, R., Xuong, N., *et al.* (1977). *Science*, **197**, 452.
58. Navia, M. A., Springer, J. P., Poe, M., Bojer, J., and Hoogsteen, K. J. (1984). *J. Biol. Chem.*, **259**, 12714.
59. O'Hara, B. P., Wood, S. P., Oliva, G., White, H. E., and Pepys, M. B. (1988). *J. Cryst. Growth*, **90**, 209.
60. Evans, P. R., Farrantes, G. W., and Lawrence, M. C. (1986). *J. Biol. Chem.*, **191**, 713.
61. Krause, K. L., Volz, K. W., and Lipscomb, W. N. (1985). *Proc. Natl. Acad. Sci. USA*, **82**, 1643.
62. Schar, H.-P., Zuber, H., and Rossman, M. G. (1982). *J. Mol. Biol.*, **154**, 349.
63. Skarzynski, T., Moody, P. C. E., and Wonacott, A. J. (1987). *J. Mol. Biol.*, **193**, 171.
64. Soderberg, B.-E., Holmgren, A., and Branden, C.-I. (1974). *J. Mol. Biol.*, **90**, 143.
65. Leslie, A. G. W., Moody, P. C. E., and Shaw, W. V. (1988). *Proc. Natl. Acad. Sci. USA*, **85**, 4133.
66. Tucker, A. D., Pattus, F., and Tsemoglou, D. (1986). *J. Mol. Biol.*, **190**, 133.
67. Ollis, D. L., Brick, P., Hamlin, R., Xuong, N. G., and Steitz, T. A. (1985). *Nature*, **313**, 762.

Chapter 4
Protein purifications from mammalian cell culture

Terry Cartwright

TCS CellWorks Ltd, Park Leys, Botolph Claydon, Buckinghamshire MK18 2LR, UK.

1 Introduction

Health care in the last decades of the twentieth century has been revolutionized by our capacity to produce recombinant proteins and viruses for therapeutic, vaccine, and diagnostic use. The greatest challenge has been the production of complex proteins for administration to humans in order to replace or augment deficient endogenous proteins. In many cases, the complexity of the protein structure involved, particularly where post-translational modifications such as glycosylation are required, is such that only mammalian cells can perform all of the synthetic steps required to produce truly authentic copies of the natural human protein. One of the major successes of the biotechnology industry is the speed with which mammalian cell culture has been developed into a viable biopharmaceutical production process from its beginnings in the 1960s. Today approximately half of the multi-billion dollar biopharmaceutical industry is based on products derived from cultured mammalian cells.

Recombinant mammalian cells are able to produce accurate copies of human proteins because the way in which they control gene expression, the synthetic and processing enzymes which they employ, and their protein folding and secretion mechanisms are all essentially identical to those of human cells. While this close mechanistic similarity favours the authenticity of the recombinant protein product, it also poses specific purification and characterization problems. Proteins derived from production cells may be immunologically cross-reactive with their human homologue and so give rise to problems of sensitization, genetic material from the cells could become active in a patient receiving the drug, and the viruses which potentially infect the production cells could also infect the patient. In addition, some production cell types may constitutively produce proteins such as growth factors or cytokines which are biologically active at extremely small doses. The need to eliminate these potential hazards imposes specific requirements on purification procedures for recombinant products from mammalian cells.

The medium required for growth and product generation can also add considerably to the difficulty of purification of the required protein product. Unlike the simple media used for microbial fermentations, mammalian cell culture medium frequently requires undefined, animal-derived nutritional supplements such as serum. This significantly increases the protein load in the feedstock and introduces additional risks of contamination by viruses or other hazardous agents.

Purification strategies for mammalian cell-derived proteins have not only to produce the required product in a reproducibly pure and characterized form but also to eliminate potentially harmful contaminants which could derive from the materials used in the manufacturing process. These include animal proteins, viruses, and DNA from the cells or culture medium, and more general contaminants such as endotoxin and other pyrogens, substances leached from chromatographic columns, or antibiotics used in cell culture.

2 Nature of recombinant proteins produced in mammalian cells

Recombinant proteins produced in mammalian cells are generally secreted into the medium as complete, correctly folded native protein which has undergone all of the requisite post-translational modifications. In general therefore, problems of correct protein conformation, such as may occur when recombinant proteins are recovered from bacterial inclusion bodies, do not arise. Also in contrast to bacterial systems, mammalian cells do not produce endotoxin and consequently, its removal is not a specific problem. However, all process steps must be designed to prevent the possible introduction of bioburden and endotoxins.

Concentration of the protein product is typically low, usually below 100 mg/litre and often considerably lower than this although high-yielding hybridoma lines may produce up to a gram of product per litre. There are some indications that post-translational modifications and the secretion process itself may become rate limiting in some cells when powerful promoters are used to drive expression and work is in progress aimed at increasing specific yield per cell by manipulation of these processes. Assuring that post-translational modification is consistent, especially in the case of glycosylation, is a critical element in process design and product characterization for proteins produced in mammalian cells since small differences in glycosylation can result in major differences in efficacy, pharmacokinetics, and tissue distribution (1).

Secretion of the product already provides an important separation of the required protein from intracellular protein and the nucleic acid of the production cells. Rupture of the cells would negate this advantage and so careful attention has to be paid to maintaining cell integrity. Mammalian cells are much more fragile than microbial cells, both to mechanical forces and to adverse culture conditions. Careful control of culture and cell separation conditions is, therefore, essential to avoid subsequent contamination of the product by intracellular proteins released by cytolysis.

The main general characteristics of recombinant proteins produced by animal cells can be summarized as follows:

- secreted into the medium in the native, correctly folded form
- post-translational modifications correctly performed
- no limit on the molecular weight of the recombinant protein
- mammalian cells do not produce endotoxin
- low product concentration
- protein load in the medium may be high
- product stream may contain viruses, DNA, or cellular protein which must be eliminated by downstream processing procedures

3 Transgenic animals

The production of recombinant human proteins in the milk of transgenic animals is a recent development which uses the protein synthetic capacity of mammalian cells without the technical difficulties and heavy capital outlay for production plant imposed by their production *in vitro*. In these animals, the gene for the protein of interest is placed under the control of a mammary gland-specific promoter such as those for the casein, β-lactoglobulin, or whey acidic protein genes and is thus secreted with other milk components. This approach is currently being evaluated commercially using goats, sheep, cows, and rabbits. A significant advantage is that yields of several tens of grams per litre of milk per day of the required protein can be achieved so that production of proteins for which large quantities are required (for example, > 100 kg/year may be needed for some therapeutic monoclonal antibodies) may become much more commercially viable using transgenic animals than using cultured mammalian cells.

Some problems do exist with this approach. For example, expression of proteins that are pharmacologically active such as insulin, cytokines, adhesion molecules, etc. may not be tolerated by the animals. On the other hand, clinically important proteins such a human serum albumin, antithrombin III, and antibodies can be produced without problems of toxicity. Downstream processing issues are essentially the same as for other mammalian cell-derived recombinant proteins but a few differences exist which will be mentioned at appropriate points in the text. Some of these are summarized in *Table 1*.

Table 1 Protein production from mammalian cell systems

Cell system	Type of product	Feedstock for downstream processing	Typical product yield
Cultured recombinant cells, e.g. CHO, BHK	Essentially any required protein	Culture supernatant	100–200 mg/litre
Cultured hybridomas	Monoclonal antibodies	Culture supernatant	Up to 1 g/litre
Mammary gland of transgenic animals	Essentially any required protein	Milk	10 g/litre+

4 Overview of extraction and purification process

Recovery of pure protein for pharmaceutical use requires a relatively standardized sequence of operational phases passing through extraction, initial fractionation, main purification, and final purification (polishing) steps. Several general considerations apply to all of these.

(a) Hygienic operation—all process steps must be capable of operation under contained conditions and should not add bioburden or endotoxin to the product stream. Regulatory approval of the process will depend on this being demonstrated. Process scale animal cell culture must be performed under sterile conditions and, unlike in laboratory cell culture, in the absence of antibiotics (because of the extreme difficulty of definitively eliminating antibiotic residues from the final product). The feedstock for purification is, therefore, sterile and has low endotoxin content.

(b) Process materials that cannot be steam-sterilized (for example, chromatographic supports) must be cleaned and depyrogenated using a validated procedure. Typical approaches to this include treatment with sanitizing solutions such as 0.2–0.5 M NaOH or acid ethanol (60% ethanol/0.5 M acetic acid). Sodium hydroxide/ethanol mixtures (e.g. 50% ethanol/0.2 M NaOH) are also used. Exposure time is typically 4–24 h at room temperature.

(c) A consequence of treatment with the sanitizing solutions mentioned above may be leaching of chromatographic ligands or other contaminants. This aspect of the process also requires qualification.

(d) Whenever possible, use of proteins of animal origin (e.g. use of antibody affinity columns) should be avoided in processes for the production of pharmaceutical proteins since regulatory authorities require that the ligand used should also be of pharmaceutical quality. This poses obvious problems of expense and practicality.

(e) Rapid reduction in volume and rapid initial clean up is highly desirable so that the later phases of purification, which tend to use more expensive chromatographic supports, are supplied with a concentrated, enriched feedstock. It is also highly desirable to separate the desired product rapidly from contaminants such as proteases or glycosidases which could degrade the product. Efficient product capture from the feedstock is the most effective way of achieving this.

(f) Process integration—it is very desirable wherever possible to ensure that the product from one process step can be directly passed to the following step without the need for further preparation, for example an additional buffer change requiring dialysis or diafiltration.

A typical purification scheme, applicable to recombinant proteins or monoclonal antibodies produced in mammalian cell culture or from transgenic animals is summarized in *Table 2*.

Table 2 Typical purification scheme

Input material	Operation	Processes	Typical product purity
Initial fractionation			
Initial feedstock, e.g. culture supernatant	Clarification (if needed)	Microfiltration Centrifugation	Up to 30–50%
Clarified supernatant	Concentration (if needed)	Ultrafiltration usually	Effectively unchanged
Main purification			
Concentrated supernatant in selected buffer	Selective capture of required product	Affinity chromatography Ion exchange Hydroxyapatite Hydrophobic binding	> 90%
Eluate from capture step	One or two further purification steps—typically chromatographic	As above	99–100%
Polishing			
Purified product	Removal of aggregates Process additives etc.	Size exclusion chromatography	99–100%
Formulation and filling			

5 Process design—preventive measures

As with many production processes, some of the most effective process design steps that can be taken are the preventive measures which ensure that the product stream to be processed is as enriched and as concentrated as possible and does not contain intractable contaminants whose removal will present major problems further downstream. In the case of products from animal cells, careful selection of the cells, the way in which they are grown, and the medium used to grow them provide important benefits in the subsequent purification.

5.1 Cell culture system

Different types of mammalian cells require different culture conditions. Some, known as anchorage-dependent cells, will only grow as monolayers attached to a suitable surface while others grow in free suspension in stirred fermenters. Stirred suspension cultures are generally easier to control and to scale-up but require an additional separation step to separate cells from the medium which contains the secreted product. In cultures of anchorage-dependent cells, the medium can simply be pumped off from the cells (although clarification to remove cell debris is usually still required).

The low concentration of product in medium from cultured recombinant mammalian cells is in part due to the low cell density that can be achieved in classical suspension or monolayer cultures (typically $1–5 \times 10^6$ cell/ml). Several systems designed to permit up to 100-fold higher cell densities are now available and are being evaluated increasingly for biopharmaceutical production. Tight control of cell culture conditions is necessary in these systems to minimize cell death and the release of cell debris into the product stream and this constitutes one of the major practical problems remaining. In some configurations, cells are kept in a compartment which is separate from the bulk culture medium, again providing a means of increasing product concentration and, to some extent, of preventing contamination of the product by medium components (2).

5.2 Production cells

Many of the cell types that are routinely used for the production of proteins (for example most hybridoma lines and the CHO cell line) show the presence of viral sequences in their genome or virus-like particles within the cell. As will be discussed later, knowledge of which viruses could be released by the cells in part determines the purification strategy.

Regulatory authorities require that mammalian cells used for production of pharmaceutical proteins be rigorously characterized in terms of their potential contamination with viruses and other infectious agents with particular reference to the possibility of latent viruses or viral genomic material becoming activated over the time for which the cells are maintained in culture for production.

The characterized cells are cryopreserved and held in liquid nitrogen as a 'Master Cell Bank' A second bank, the 'Working Cell Bank' is produced from a few ampoules from this. Each production run is then begun with a fresh ampoule of cells from the Working Cell Bank. In addition to the biological safety aspects, this approach also contributes greatly to the reproducibility of production batches.

5.3 Culture medium

The medium used is of critical importance to the strategy applied to purification of recombinant protein products. Mammalian cell culture medium is always complex, containing salts, vitamins, amino acids, carbohydrates, and a variety of additives designed to improve cell growth and/or product yield. These include serum (usually fetal calf serum, FCS) frequently used at 10% (v/v) concentration but sometimes combinations of other proteins including albumin, transferrin, and insulin are used to replace serum in order to reduce the protein load and to provide a more 'defined' medium. Manufacturers have made major efforts to develop media which contain no exogenous proteins and this has been achieved for some of the less fastidious hybridomas and a few other cell types (3). However, use of serum-containing medium is still widespread in the biopharmaceutical industry and poses several practical downstream processing problems as follows:

(a) High level of contaminating proteins (10% FCS contributes ca. 4 g of protein per litre!).

(b) Significant batch-to-batch variation occurs in serum performance and composition.

(c) Possibility of contamination of the product with serum-derived viruses and other agents such as BSE.

(d) High cost (typically 10–20 × the cost of all other medium components!).

It is practice to screen serum batches for the presence of viruses before use but, as the current BSE phenomenon has illustrated very clearly, the presence of unknown infectious agents may have very major consequences and would not be protected against by screening. Clearance of potential infectious agents by the purification process remains the major safeguard as will be discussed in later sections.

Physicochemical similarity between medium protein constituents and the required product may also cause problems. Animal immunoglobulins present in serum-containing medium used for the production of monoclonal antibodies from hybridoma culture can be difficult to separate efficiently from the desired immunoglobulin product ('immunoglobulin-stripped' serum is available commercially to address this problem). Transferrin may also be difficult to remove from antibody preparations since it shows similar charge properties to some Ig subclasses.

Another inconvenience of standard tissue culture medium is that the phenol red, which is usually included as a pH indicator, may cause problems by binding irreversibly to some chromatographic media used in purification.

6 Cell separation

When cells cultured in suspension are used, it is generally necessary to separate the cells from the product-containing medium. This must be achieved with the minimum of cell damage or lysis since this would result in additional contamination of the product stream as discussed earlier. It is also very desirable to be able to retain viable, undamaged cells for future product generation, particularly in the case of perfusion systems where fresh medium is fed to the cells and product removed on a continuous or semi-continuous basis.

Unfortunately, animal cells are very susceptible to damage by mechanical shear forces which means that only very gentle separation procedures can be applied. Standard tangential flow filtration systems, which require high flow rates to prevent membrane fouling, and conventional centrifugal separators are widely used but often yield a significant proportion of damaged cells. In fact, mammalian cells sediment easily (some processes use initial sedimentation under gravity) and can be very efficiently separated at say $1000–1500\ g$ without damage. A recent development has been the design specifically for animal cell separation of totally enclosed centrifugal separators which use low g force, low shear peristaltic pumps, and no rotating seals. One such device, the Centritech® developed by Sorvall, can be operated at flow rates of 100 litres/hour. If needed, final clarification of the medium after cell separation can be performed either by centrifugation or by micro-filtration, both of which can be operated at high flux under contained conditions.

Another promising new approach is the so-called 'acoustic filter' where cells are concentrated at the velocity antinodes of ultrasonic standing waves. Efficient laboratory scale systems have been produced and closed, sterile operation can be readily achieved (4). At present, heat generation in the separation chamber is the limiting factor for application of this technology to the separation of animal cells and successful process scale systems have not been so far produced.

6.1 Purification without cell separation

Recently, several chromatographic media have been introduced which can effectively capture the required product directly from the cell culture without the need for prior separation of the cells. These will be discussed in more detail later in this chapter. An important point to keep in mind when using these is that optimal binding conditions for the product are unlikely to be identical to the conditions required for optimal cell survival, and that small deviations from the latter may result in cell lysis and release of cell contents and so increase the burden of contaminants needing to be eliminated further downstream.

7 Initial product recovery and fractionation

Because cell lysis is minimal, mammalian cell culture supernatants do not suffer from the sort of heavy loading with nucleic acid which may occur with bacterial cell lysates. Nuclease treatment to reduce viscosity of the feedstock is therefore unnecessary.

The clarified culture fluid is relatively low in protein concentration (certainly in concentration of the required protein) but contains a high concentration of low molecular weight components from the culture medium (ionic strength will be close to 0.15). The initial purification is therefore aimed at reducing the major contaminant (water) and producing a feedstock in an appropriate buffer for subsequent purification steps. Some purification with respect to contaminating proteins can also be achieved.

Methods which can achieve these objectives include protein precipitation (by salt or by solvents), adsorption of the required protein, and aqueous liquid/liquid extraction (5). Precipitation methods are not widely used with feedstocks of low protein concentration since precipitate formation is often not efficient in these conditions and also large quantities of precipitant are required. Ultrafiltration procedures are now almost universally used for this operation since they:

- are easy to operate under contained conditions
- do not subject the protein to extreme or denaturing conditions
- are easy to scale-up if required
- can simultaneously concentrate the protein and change the buffer in preparation for the next purification operation
- make no addition of potentially contaminating substances to the product stream

A wide range of sterilizable or sanitizable ultrafiltration systems at all scales are now available from all of the major manufacturers. In some situations, where the required protein differs significantly in size from contaminating proteins, useful purification may also be achieved. However, ultrafiltration is not very resolutive owing to the high bandwidth of particles that will pass the membrane. It is now usual to use a cut-off of say 10 kDa and retain essentially all of the contaminating proteins with the product. It should be noted that this approach would also concentrate any virus particles which might be present.

One limitation of ultrafiltration systems is that they may be subject to membrane fouling by particulates or by formation of a gel layer on the membrane by some proteins. Some antifoams used in mammalian cell culture may also foul ultrafiltration membranes. This tendency is minimized if the feedstock is properly clarified and ultrafiltration systems with appropriate membrane clearing hydrodynamic characteristics are used.

7.1 Recovery of recombinant protein from milk

Milk is a complex colloidal suspension containing high concentrations of casein micelles and other milk proteins, fat globules, host cells, and proteins and nucleic acids from lysed host cells. Initial product recovery and fractionation uses techniques established in the dairy industry and includes fat removal by skimming and centrifugation, acid or salt precipitation of casein, and filtration to remove particulates. These additional steps add significantly to the cost of producing pure proteins from milk and the result is a rather lower percentage yield of the final product than would be typically obtained from a cell culture supernatant. The clear whey fraction containing the protein of interest then serves as a concentrated feedstock for subsequent purification steps.

Another point that may complicate recovery of the required pure protein from milk is the variability of the feedstock which may arise due for example to variations in animal feed or in the hormonal status of the animals. This is not generally considered to be a major problem.

7.2 Direct capture of product from feedstock

Direct selective capture of the product from the feedstock is a very desirable first step in any purification procedure and, as will be discussed later, affinity systems lend themselves particularly well to this. This approach is particularly advantageous if binding will occur without modification of the feedstock solvent conditions since it may then be possible to eliminate the need for an earlier concentration and/or buffer change step and to feed clarified culture supernatant or whey proteins directly to the affinity support with corresponding savings in time and material. As mentioned in Section 6.1, several chromatographic media have been recently introduced which are tolerant of particulate material in the feedstock and for which even the clarification step may not be needed.

8 Main purification

Proteins are produced from mammalian cells predominantly for biopharmaceutical applications. The task of the main purification process for these high value products is to produce a protein that is safe, pure, and reproducible. It is important to achieve this in as few process steps as possible in order to maintain high yields, to enhance the robustness of the process, and to optimize process economics. This requires the selection of a series of complementary unit processes of the highest possible resolving power to give the necessary protein purity and, specific for products derived from animal cells, to achieve satisfactory clearance of possible viral and DNA contaminants. Most commercially viable processes achieve this in three to five purification steps. Many techniques are available for this and a full discussion of these is beyond the scope of this article. Practically, column chromatography methods predominate and these will be considered briefly with special reference to mammalian cell products.

8.1 Ion exchange chromatography

This is the most commonly used procedure at present. A wide variety of different column support matrices are available and many processes are based on several successive ion exchange separations performed under different conditions of pH and/or ionic strength. One possible practical limitation of the procedure is the need to feed material to the column at low ionic strength which probably imposes the need for a buffer change between steps. This can be conveniently achieved by diafiltration. A rapidly growing area is the development of ion exchange supports optimized for the capture of the required product direct from the feedstock.

8.2 Hydrophobic interaction chromatography

This approach relies on the binding of hydrophobic groups on the protein surface to a hydrophobic support phase. Binding is strongest at high ionic strength and selective elution of proteins can be achieved by reducing ionic strength or by changing the polarity of the mobile phase by addition of an organic solvent (reverse-phase chromatography). This latter approach is very successful with peptides or with some of the rather small, thermodynamically very stable proteins such as cytokines and interferons which were among the first targets for recombinant protein production. However, the technique is less successful with other proteins where exposure to organic solvent may result in an irreversible partial unfolding of the protein. Such partially denatured proteins may be very difficult both to detect and to eliminate further downstream.

A potential advantage of the different conditions needed for loading columns for these two types of chromatography is that it may be very practical to use them in tandem since the high ionic strength eluate from an ion exchange column may be appropriate for direct application to a hydrophobic interaction chromatography column and vice versa.

Both of the above techniques require the binding of a mixture of proteins to the column followed by selective elution of different protein fractions by changing the characteristics of the mobile phase either stepwise or using a gradient. It follows, therefore, that the resolutive power of the system depends heavily on the quality of the column stationary phase. A great deal of development effort has been put into this, trying simultaneously to maximize column binding capacity, flow rate, and tolerance to particulates, and to improve resolving power—all aspects which augment unit process efficiency. These often mutually competing aims have been largely achieved in some of the new supports now available. The most advanced column supports are correspondingly expensive, but the high value of biopharmaceutical products makes this acceptable.

8.3 Affinity chromatography systems

Affinity chromatography is potentially the most powerful and selective approach to protein purification. The technique relies on the ability of proteins to bind selectively and reversibly to specific ligands immobilized on a support matrix. Unlike the methods described above, affinity columns rely on specificity at the adsorption stage to achieve high resolution. In the ideal case, where the protein required binds with absolute specificity to a ligand, all other proteins can be washed from the column and the desired protein eluted resulting in single step purification. There are several systems reported where this ideal is close to being achieved. In other cases, lack of binding specificity or the harsh conditions needed for elution may limit its usefulness.

8.3.1 Protein affinity ligands

Absolute specificity can be achieved in some cases, for example when an immobilized monoclonal antibody is used to purify its antigen by 'immunoaffinity' or when an immobilized hormone is used to purify its receptor. Single step immunopurification can be achieved with elution by lowered pH, but, with antibodies of high affinity, elution of the antigen without damage by exposure to pH extremes may be a major problem. In such cases, use as a 'competing ligand' of a peptide containing the sequence recognized by the antibody may permit elution under milder conditions.

The major limitation of this approach for the production of biopharmaceuticals is the regulatory concern associated with the use of a second animal-derived protein in the purification system. Thus the immobilized antibody itself is perceived as constituting an additional risk of contamination by viruses or other agents and it is therefore required that the antibody used should itself be of pharmaceutical quality. This approach then becomes prohibitively expensive.

8.3.2 Protein A and protein G

In the specific area of antibody purification, protein A (from *Streptococcus aureus*) has been widely and successfully used. This protein binds with high affinity and specificity to the Fc region of most immunoglobulin subtypes. Elution can be achieved by lowering the pH. Some immunoglobulin subtypes do not bind adequately to protein A and for these, a second bacterial protein, protein G has proved effective. Both of these proteins are available as highly purified recombinant products bound to a number of suitable column matrices and are widely employed in antibody purification at laboratory and industrial scale. A useful observation for those producing monoclonal antibodies in transgenic goats is that goat IgG does not bind to some protein A supports (6). Protein L is another bacterial antibody binding protein which has been recently introduced. Unlike proteins A and G, protein L binds antibody via the κ light chain and can consequently be used for purification of a wider range of Ig subclasses (7). Protein L is commercialized by Actigen Ltd.

8.3.3 Lectins

The lectins are a large family of proteins (mainly from plant sources) which selectively bind the different sugar moieties which occur in glycoproteins. While not absolutely specific for a given protein, lectin affinity chromatography can result in very useful purification in a single step. Desorption from lectins can be achieved by altering pH but elution under mild conditions is frequently achieved by using the relevant free sugar as a competing ligand. With all protein affinity ligands, it is of critical importance to ensure that leached ligand does not appear as a contaminant in the final product. This is particularly important for lectins which are frequently toxic, behaving as potent haemagglutinins or mitogens.

8.4 Small molecule affinity ligands

In some cases, small molecules may exhibit a strong and specific affinity for proteins. Examples include the affinity of substrate molecules for the active site of their enzyme or for some other functional site on a protein (such as the lysine binding site on tissue plasminogen activator or the nucleotide-binding site on numerous regulatory proteins). In these cases, elution from the column can frequently be achieved using the substrate or a substrate analogue as a competing ligand.

Where such affinity binding sites do not exist in a recombinant protein, they can be built in at the genetic manipulation stage to give a so-called 'affinity tag' (8). This approach has the advantage that a number of different recombinant proteins can all be given the same affinity tag and can all therefore be engineered to give a good, single step purification using the same column system. Frequently used affinity tags include polyhistidine tags which permit purification by metal chelate chromatography and the creation of glutathione-S-transferase fusion proteins which can be purified on immobilized glutathione. A key consideration with all such fusion proteins is the need to be able to cleave off the tag without damaging the required protein, and to unequivocally separate the protein from the uncleaved fusion protein and from all cleavage products.

8.4.1 Pseudo-affinity systems

This is the name given to artificially created affinity ligands where a protein is empirically determined to bind to an immobilized small molecule with a greater or lesser degree of specificity. The use of small molecule ligands is generally easier to control and to monitor than that of proteins and any leached ligand may be easier to eliminate from the final product.

The most commonly used pseudo-affinity ligands are a series of triazine dyes bound to suitable matrices. The prototype for these was Cibacron Blue F3GA which was discovered serendipitously to bind proteins when it was used coupled to high molecular weight dextran for determination of the void volume of size exclusion chromatography columns. A wide variety of 'second generation' dye ligands have now been developed, both to add to the repertoire of available ligands and to overcome the problem of leaching of the dye under conditions of extreme pH which was observed with some of the earlier dye affinity supports. The latest dye affinity columns will withstand treatment with sanitizing solutions, including sodium hydroxide, without leakage. Dye affinity separations are growing rapidly in popularity since:

- protein binding capacity can be very high
- appropriate dyes are readily available cheaply and in quantity
- coupling can be performed simply with non-toxic reagents
- ligand leaching is minimal when appropriate matrix is used
- high resolution can frequently be achieved

8.4.2 Development of new affinity ligands

The observation that many small molecules can bind to proteins with sufficient specificity and affinity to permit efficient chromatographic separation has led to an intensified search for new pseudo-affinity ligands. Some companies (for example Affinity Chromatography Ltd.) have evaluated a wide range of existing dye molecules for this purpose. Some of the most recent advances in this area have harnessed the power of combinatorial approaches to permit rapid screening and identification of new ligands. Thus, Dyax Corporation is using phage display technology to discover new small peptide ligands and Affinity Chromatography Ltd. is screening combinatorial libraries of synthetic molecules as potential affinity ligands. The rapidly developing versatility and robustness of pseudo-affinity supports suggests that this may become the dominant separation technology for recombinant proteins in the near future.

9 Particle-tolerant chromatographic supports

Chromatographic supports that can be used without prior clarification of the feedstock have already been mentioned at several points. Examples of this new generation of robust stationary phases include the PROSEP® range from Bioprocessing Ltd. which is based on a rigid glass bead support system and the STREAMLINE® range from Pharmacia which uses macroporous beads of cross-linked agarose around a quartz core. Both of these are very stable adsorbents which are resistant to most of the cleaning and sanitizing reagents in common use, although the use of sodium hydroxide is not recommended with PROSEP®-

A. Both also have the advantage of being usable in a fluidized bed-type configuration in which crude feedstock containing cells and cell debris can be fed directly to the absorbent without significant loss in the efficiency of product capture or resolution. Cells and debris can be easily washed out of the column along with unbound protein.

Another interesting recent development is the CELLTHRU BIGBEAD ion exchange resins from Sterogene Bioseparations. These products can also accept cell suspensions without pre-treatment, this time in a packed bed format due to the large bead size used which allows uninterrupted passage of suspended solids in the feed stock. CELLTHRU BIGBEAD can also be effectively cleaned in place using NaOH and all other sanitizing solutions in common use.

A general characteristic of ion exchange-based systems is the need to alter the pH and ionic composition of the feedstock somewhat from the cell culture medium (typically pH 7.2–7.5, ionic strength 0.15) in order to produce suitable conditions for loading the resin. This can be achieved by diafiltration or dialysis or, quicker and simpler, by dilution (as is used in *Protocol 2*). Sterogene has developed a specific resin (IonClear BigBead™) which reduces ionic strength by capture of both anions and cations with negligible adsorption of protein. The resin also removes phenol red from culture medium, can be effectively cleaned using the usual cleaning agents, and can be autoclaved.

10 Purification of monoclonal antibodies

Monoclonal antibodies as a class certainly represent the group of proteins for which the most industrial experience has been accumulated and illustrate well the problems encountered in the production of pure protein products from mammalian cells. The continuing use of monoclonal antibodies in the diagnostic field and the growing need for 100 kg scale production for therapeutic purposes has tested the technology and economics of protein purification to its limits. The overall similarity of antibody molecules has allowed 'generic' purification systems to become established where the same column system can be used for several products. However, some process development is still required to optimize purification of a given antibody.

Monoclonal antibodies were formerly widely produced in ascitic fluid in mice but for practical and ethical reasons, this procedure was effectively superseded by production using cultured mammalian cells. With the need for increasing quantity, production in cultured cells is itself coming under threat from production in the milk of transgenic animals.

10.1 Separation using protein A

Protein A affinity chromatography has been widely used for antibody purification and in some instances is almost ideal, giving antibody purity in excess of 95% at high yield in a single step. Protein A chromatography is highly effective in the removal of endotoxin, DNA, and viruses. However, there are practical limitations because of its inefficient binding of some immunoglobulin isoforms and the need sometimes for harsh conditions to achieve elution which can both damage the antibody and result in leaching of protein A from the column. This is particularly important since protein A has been shown to have immuno-modulatory activity *in vivo*.

Nevertheless, the most recent developments of protein A supports have addressed most of these problems by offering improved immobilization chemistry and optimal protein A orientation on the column to widen the range of Ig subclasses bound. Most recently, protein A column bed materials that permit the application of unclarified culture medium to the column without any pre-treatment such as dialysis, concentration, pH change, or filtration have been developed.

Whichever variation of the protein A (or protein G) system is used, the procedure is basically similar to the example shown in *Protocol 1*. In all cases, elution is performed using a low pH buffer which may include other reagents such as chaotropic salts to facilitate desorption. For most antibodies, it is essential to rapidly return the pH to near neutrality and to dilute out the effects of any other reagents used to avoid further damage to the antibody. This may be used with advantage to prepare the protein for the next step.

Protocol 1

Capture of monoclonal antibody from tissue culture supernatant using a protein A adsorbent (PROSEP®-A)

Equipment and reagents

- Suitable column chromatography equipment
- 0.2 or 0.45 μm filters
- Binding buffer: 25 mM Tris–HCl, 150 mM NaCl pH 7.4

- Elution buffer: 0.1 M citrate pH 3.0
- Neutralizing buffer: 1 M Tris–HCl pH 9.0
- Regeneration buffer: phosphoric acid pH 1.5 (approx. 10 ml phosphoric acid in 1 litre distilled water)

Method

1 Prepare the packed column in accordance with the manufacturer's instructions. Wash with 5–10 column volumes of binding buffer at the required flow rate (about 80 ml/min for a 5 cm diameter column).

2 Clarify the tissue culture supernatant by filtration through a 0.2 or 0.45 μm filter (not necessary if operating in fluidized bed mode).

3 Apply the sample to the column. Approximate binding capacity is 40 mg Ig/ml but this should be optimized by experiment.

4 Wash with 5–10 column volumes of binding buffer (or until OD of washings comes down to zero).[a]

5 Elute product by passing elution buffer at a flow rate of approx. 50 column volumes/h.

6 Collect fractions directly into neutralizing buffer to avoid damage to the product by exposure to acid pH.

7 Regenerate the column by passing regeneration buffer until no protein is detectable in the effluent. Re-equilibrate the column with binding buffer as in step 1.

8 The column can be sanitized using guanidine or other protein solubilizing agents. The column can also be treated with ethanol/acetic acid mixtures. Treatment with sodium hydroxide is not recommended.

[a] Contaminating bovine immunoglobulins from tissue culture medium can be eliminated by careful adjustment of washing conditions at this stage. This has to be done by experiment on a case-by-case basis.

10.2 Monoclonal antibody purification by ion exchange

The disadvantages of protein A and related systems (cost, leaching, harsh elution conditions, and ineffectiveness with some Ig subclasses) have lead several laboratories to evaluate alternative approaches and this has resulted in the production of a number of (mainly) cation exchange systems that are specifically geared to the capture and purification of monoclonal antibodies. Again emphasis has been on high performance supports with high capacity, good chemical and mechanical stability, and the ability to capture antibody under conditions of pH and ionic strength that are not far removed from those of the feedstock and which do not require harsh conditions for elution. Examples include STREAM-LINE supports from Pharmacia, Bio-Scale S2 from Bio-Rad, Bakerbond ABx from J. T. Baker, and CELLTHRU BIGBEAD ion exchange resins from Sterogene Bio-separations. As before, the STREAMLINE supports can handle unclarified supernatants and cell suspensions when used in a fluidized bed mode.

Protocol 2

Capture of monoclonal antibody from clarified culture supernatant using a 'mixed mode' ion exchange medium (Bakerbond™ ABx)

Equipment and reagents

- Suitable column chromatography system
- Buffer A: 10 mM 2(N-morpholino)-ethane sulfonic acid (MES) pH 5.6
- Buffer B: 500 mM sodium acetate pH 7.0 or 500 mM $(NH_4)_2SO_4$ in 10 mM sodium acetate pH 5.6
- Bakerbond™ ABx 40 μm size, prepared according to the manufacturer's instructions

Method

1 Pack a suitable sized column and equilibrate with buffer A by passing about 15 column volumes at a flow rate of about 90 ml/min for a column of 5 cm i.d. (recommended flow rates for different column dimensions are given in the manufacturer's literature). The pH of the effluent should be 5.6 when the column is fully equilibrated.

2 Dilute the cell culture supernatant to be loaded with four volumes of buffer A.

3 Load the sample in buffer A at a flow rate 70% of that used for washing in step 1. Column capacity is about 150 mg of total Ig per gram of ΛBx.[a] All immunoglobulin isotypes bind effectively.

4 Wash the column with one column volume of buffer A.

5 Develop the column using a linear gradient from 0% buffer B to 100% buffer B over 1 h at the flow rate used in step 1.[b]

6 Regenerate the column by passing 20 column volumes of high ionic strength salt solution such as 2 M sodium acetate or 1 M potassium phosphate buffer at pH 7–8.

7 If necessary, clean the column using protein solubilizing agents such as guanidine or detergents (see manufacturer's literature).

8 Sanitize the column if required using ethanol or aqueous solution of bactericide. Short treatment with 0.1 M sodium hydroxide is possible but is not recommended.

[a] The column binds contaminating antibody in serum-containing medium as well as the required antibody. Thus binding capacity for the required antibody is higher if serum-free medium has been used for production.

[b] Fractionation of different Ig species occurs during the gradient and this can be used to separate contaminating immunoglobulins from the required product. Some process development may be required to optimize this.

10.3 Other recent approaches to purification of monoclonal antibodies

Immobilized metal affinity chromatography (IMAC) (8) has also been found to be very effective for the purification of immunoglobulins. These bind because of the presence of a highly conserved histidine cluster which occurs in the Cg3 region of all mammalian immunoglobulins. Capture is reported to be efficient with elution being readily achieved under mild conditions.

Hydroxyapatite has also recently been described for purification of monoclonal antibodies. This highly selective chromatographic support has become practical for process scale operation with the introduction of new ceramic macroporous hydroxyapatite (e.g. Bio-Scale CHT1–2 from Bio-Rad).

11 The elimination of viral contamination risks

Possible contamination by retroviruses or other viruses remains the most serious safety concern for the products of mammalian cells. Retroviruses represent a particular problem since they are associated with cancer and immunosuppressive diseases, they frequently persist as latent infections not recognizable by standard virus detection techniques, they become integrated into the genome and can be inherited as a DNA provirus, and their host range and virulence can alter spontaneously due to recombination between exogenous retrovirus and proviral sequences. Several herpes viruses are also of concern because of their widespread occurrence in human and animal populations, their propensity to persist as latent infections, and their association with cancer and other serious diseases.

As mentioned earlier, the risks of viral contamination are partially addressed by careful screening by virus infectivity assays of the producer cells which go to make up the cell banks. In addition, regulatory authorities require that representative cells (the 'end-of-production cells' or EPC) should be cultured beyond the point at which production normally takes place and again tested for virus content by a range of sensitive methods. This approach is intended to identify situations in which latent virus infection in the cells could become activated during the period of cell culture. Any virus detected at this stage must be quantified and if possible identified to establish the extent of virus clearance for that particular virus type which the purification process will have to achieve. However, even if infective virus is not detected in any of these tests, this cannot preclude the presence of unknown agents for which no test is currently applied. Also, although the producer cells may not appear to contain active virus, most do contain virus-like particles visible under the electron microscope suggesting that the potential to produce live virus may exist.

In addition to these screening procedures-based infectivity assays, the purification process is required to provide definitive protection against viral contamination. Purification protocols are required to contain at least two 'robust' virus inactivation or removal steps (by which is meant steps whose capacity to eliminate virus has been proven in a wide variety of experimental conditions) which function by different mechanisms. There is also a requirement to demonstrate by properly validated virus clearance studies that the purification is able to clear virus effectively (9). Generally, it is not necessary to include specific virus removal steps in the downstream processing protocol since many of the purification procedures used will also reduce virus titre very effectively. Also, effective virus inactivation can be achieved by exposure to non-physiological pH, ionic strength, solvents, or temperature. The virus clearing capacity of the overall downstream processing sequence can be demonstrated by validation of the clearance efficiency of the individual processing steps.

11.1 'Spiking' and the use of model viruses

Given that the cleanest possible producer cells are used to generate recombinant proteins, the virus loads likely to be encountered in real purification runs

Table 3 Recommended model viruses for clearance studies

Virus	Family	Size (nm)	RNA/DNA	Envelope	Stability[a]
Canine parvovirus	Parvovirus	22	DNA	No	High
Polio	Picornavirus	25	RNA	No	Medium
Reovirus	Reovirus	60	RNA	No	High
SV40	Papovirus	45	DNA	No	High
Murine leukaemia	Retrovirus	80	RNA	Yes	Low
Adeno	Adenovirus	70	DNA	No	Medium
HIV	Retrovirus	90	RNA	Yes	Low
VSV	Rhabdovirus	80	RNA	Yes	Low
Parainfluenza	Paramyxovirus	150	RNA	Yes	Low
IBR	Herpesvirus	100	DNA	Yes	Medium
Pseudorabies	Herpesvirus	120	DNA	Yes	Medium
Vaccinia	Poxvirus	600	DNA	Yes	High

[a] Ability to persist in viable form under conditions likely to be encountered during protein purification

are usually below the limits of reliable detection. To measure virus clearance meaningfully it is practice to use 'spiking' experiments where large quantities of an infectious virus are deliberately added prior to a purification step so that an accurate clearance factor can be determined. For reasons of safety and convenience, these tests are performed using 'model viruses', that is viruses that are closely related to the potentially contaminating pathogenic virus of concern. One practical reason for this is that it is necessary to have virus available of very high titre to permit the measurement of large clearance factors (these are frequently of the order of 10^4 to 10^5 fold). Many of the 'real' potential contaminants identified from screening of the production cells cannot be practically or safely grown to such high titres. To cover the risk of possible unknown viruses being present it is necessary to validate the clearance of representative samples from all of the major classes of virus: RNA and DNA, large and small, enveloped and non-enveloped, etc. (see *Table 3*).

11.2 Scale-down of unit operations

Classically, validation procedures are performed using identical plant to that used on the manufacturing scale. This criterion cannot be applied to virus clearance validation because of the extreme technical difficulty and expense involved in producing high titre virus on the requisite scale and the risk of unrecoverable contamination of the plant. In addition, the safety hazards posed by large doses of viruses, even with supposedly 'harmless' viruses, are effectively unknown.

This issue is addressed by employing small scale models (of say 1/500 or 1/1000 scale) in which all the process parameters are kept identical to those in the full-scale plant. This includes factors such as column heights, linear flow rates, column residence times, buffer conditions, temperature, pH, etc. (10). All filtration membranes and column support matrices used in the model must also be identical to those used in the full-size process.

An important point to consider is that the virus spike itself must not cause perturbation of the system for example by significantly changing overall protein loading or the volume of sample applied. This can usefully be verified by ensuring that product elution characteristics are unaltered by addition of the spike and provides an additional reason for using purified virus preparations of the highest possible titre for this purpose.

11.3 Calculation of virus clearance

The virus clearance achieved should be reported as the 'reduction factor' which is the ratio of the $\log_{10}$ of the virus load (titre $\times$ volume) in the spiked starting material for a given step to that in the product from that step. Reduction factors obtained in sequential steps can be summed to calculate an overall reduction factor for the process, providing that the individual steps are independent and achieve separation according to different physicochemical parameters. Reduction factors achieved by repeated application of the same separation step are unlikely to be additive. It is also important to ensure that the procedures used are statistically valid. Thus the starting virus titre must be sufficiently high to permit accurate determination of the reduction factor and the samples of product tested must be sufficiently large that there is a high probability of detecting virus if present (12). Because of the inherent inaccuracy of virus titrations, it is practice to consider clearance factors below one log to be insignificant. *Table 4* gives an indication of the reduction factors that may be expected for different viruses in various process steps. However this is only a general guide and clearance must, of course, be validated for the specific system being used.

11.4 A cautionary note

It must not be forgotten that, as well as clearing virus, chromatographic columns can accumulate and concentrate virus which may then be released at a

Table 4 Typical titre reduction factors[a] which might be expected for several viruses of concern[b] after treatment with various process steps commonly applied to purification of recombinant proteins and monoclonal antibodies

Process step	Poliovirus	Xenotropic murine leukaemia virus	Herpes virus
Anion exchange	3–4 logs	Over 4 logs	Over 5 logs
Protein A affinity separation	3–4 logs	3–4 logs	Over 6 logs
Size exclusion chromatography	Over 3 logs	Over 3 logs	Over 3 logs
Acid treatment (pH 3.5 or below)	<1 log	Over 4 logs	Over 3 logs
Filtration (35 nm pore size)	<1 log	5–7 logs	6–8 logs

[a] $\log_{10}$ (volume $\times$ virus titre of input material) / (volume $\times$ virus titre of product).

[b] Xenotropic murine leukaemia is a retrovirus frequently used to model the retrovirus potentially present in murine hybridomas. Herpes virus is one of the virus types of particular concern with human- or primate-derived production cells.

later stage. This possibility, and the ability of the column cleaning procedures to eliminate all virus between runs, must also be validated as must the number of cycles for which a given step can be expected to achieve the required virus reduction factor (12).

11.5 Virus removal by membrane filtration

There are some situations in which it may be necessary to introduce a further process step specifically for further virus elimination and several companies have produced filtration membranes specifically for this purpose. These filters will produce significant clearance of virus with very little effect on the yield of the required protein product. The PLANOVA membranes produced by the Asahi Chemical Industry Co. Ltd. represent one of the most developed of these. PLANOVA membranes are produced with pore sizes of 15, 35, and 75 nm and have been shown to give reduction factors of 4–6 logs with model viruses that are larger than the mean pore size of the membrane (see *Tables 3* and *4*). Unlike the situation with column separations, repeated filtration cycles through these membranes are reported to give additive virus reduction factors (13).

12 Elimination of contaminating DNA

It has been shown that tumours can be produced in experimental animals by injection of pure DNA containing viral oncogene sequences and it is known that many of the cell lines used to produce recombinant proteins and monoclonal antibodies contain active oncogenes (14). There is, therefore, a risk that oncogene-containing DNA from a producer cell line persisting in a biopharmaceutical preparation could have oncogenic effects in patients. The level of risk posed by this has been the subject of much recent debate and, following WHO guidelines published in 1986, a level of 100 pg per dose has been generally considered as a level which constitutes negligible risk to patients and hence as a suitable target for purification procedures.

Recent studies have shown that earlier concerns were unduly pessimistic and very much higher doses can be injected into primates without any effect. It is probable that the limit will be revised upwards, perhaps to 500 ng per dose, in the near future. However, at present it is probably still prudent to aim for the lower limit wherever this is feasible.

The approach to the elimination of DNA is similar to that for viruses. Most protein purification procedures will readily separate protein from DNA and ion exchange steps are, particularly, effective for achieving this. As with viruses, DNA clearance validation involves spiking material to be purified with known quantities of DNA and measuring clearance in a scaled-down model of the purification system in order to obtain a DNA reduction factor. As with virus clearance studies, care must be taken that the DNA load in the spike does not perturb the performance characteristics of the purification step being validated.

A variety of DNA assay procedures sensitive in the picogram range are available. One of the most practical is the Threshold Total DNA Assay from Molecular

Devices which captures DNA on a membrane with subsequent enzymic detection of the immobilized complex (see ref. 15 for an evaluation of this procedure). More recently, direct fluorescence DNA quantitation methods have been developed (16).

13 Other biologically active contaminants

The situation can arise where the production cells used endogenously produce pharmacologically active agents such as cytokines which could not be tolerated in the final product. Leakage of material such as protein A from column supports would create a similar problem. Detection of very low levels of such agents is very difficult and spiking procedures as above are used to validate the efficacy of purification steps in clearing them from the preparation (12). Appropriately validated clearance studies may often be an acceptable alternative to routine testing of product for contamination (11).

14 Polishing steps

The final step of the purification is frequently a polishing step applied when the protein is already essentially pure and in concentrated form. Size exclusion chromatography is commonly used at this stage and is particularly useful to separate the protein from any aggregates that may have formed, to remove residual process additives, and to change the protein into the final buffer formulation required. The low flow rates achievable and the low sample volumes that can be applied to size exclusion chromatography columns militate against their use earlier in the purification sequence.

15 Problems

A significant problem with biopharmaceutical products produced from mammalian cells is the removal of endogenous or cognate proteins, that is the host animal homologue of the required human recombinant protein. These may derive from several sources including host cells and animal proteins used in culture medium in the case of cultured cells or host cells and serum proteins in the case of transgenic animals. Finely-tuned separation conditions may be necessary to remove contaminants such as endogenous immunoglobulins from monoclonal antibody preparations. As discussed in Section 5, it is best to avoid this problem by appropriate attention at the outset to such items as the medium used if this is at all possible.

Detection of trace quantities of proteins of animal origin, including cognate proteins, has most often been approached by the development of sensitive immunological tests such as that recently developed using the Threshold™ system from Molecular Devices Inc. (16).

In the transgenic animal field, one potential use of cloned animals might be to produce animals which lack their natural cognate protein to avoid later technical difficulties of efficient separation of this from the required human recombinant form.

References

1. Wright, A. and Morrison, S. R. (1997). *Trends Biotechnol.*, **15**, 26.
2. Cartwright, T. (1994). *Animal cells as bioreactors*. Cambridge University Press, Cambridge.
3. Zang, M., Trautmann, H., and Gondor, C. (1995). *Bio/Technology*, **13**, 389.
4. Trampler, F. (1994). *Bio/Technology*, **12**, 281.
5. Rosevear, A. and Lamb, C. (1988). In *Animal cell biotechnology* (ed. R. E. Spier and J. B. Griffiths), p. 394. Academic Press, London.
6. Young, M. W., Okita, W. B., Brown, E. M., and Curling, J. M. (1997). *Biopharm*, **10**, 34.
7. Nilson, B. H. K., Logdeberg, L., and Akerstrom, B. (1993). *J. Immunol. Methods*, **164**, 33.
8. Sassenfeld, H. M. (1990). *Trends Biotechnol.*, **8**, 88.
9. (1997) CBER, FDA, February 1997. Points to consider in the manufacture and testing of monoclonal antibody products for human use.
10. CPMP/268/95 (European Medicines Evaluation Agency) (1995). Revised CPMP guideline on virus validation studies. Canary Wharf, London, UK.
11. Baker, R. M., Brady, A. M., Davis, J. M., Coombridge, B. S., Ejm, L. J., and Kingsland, S. L. (1995). In *Animal cell technology: developments towards the 21st Century* (ed. E. C. Beuvery, J. B. Griffiths, and W. P. Zeijlemaker), p. 529. Kluwer Academic Publishers, Dordrecht.
12. Nakano, H., Ishikawa, G., Fuita, S., Sato, T., Manabe, S., Losikoff, A. M. *et al.* (1995). In *Animal cell technology: developments towards the 21st Century* (ed. E. C. Beuvery, J. B. Griffiths, and W. P. Zeijlemaker), p. 597. Kluwer Academic Publishers, Dordrecht.
13. Anderson, K. P., Lie, Y. S., Low, S. R., Keller, G.-A., and Dinowitz, M. (1991). *Virology*, **181**, 305.
14. McKnabb, S., Rupp, R., and Tedesco, J. L. (1986). *Bio/Technology*, **7**, 343.
15. Bolger, R., Lenoch, F., Allen, E., Meiklehohn, B., and Burke, T. (1997) *BioTechniques*, **23**, 532.
16. Ghobrial, I. A., Wong, D. T., and Sharma, B. G. (1997). *BioPharm*, **10**, 42.

Chapter 5
Protein purification from microbial cell culture

Peter R. Levison

Whatman International Ltd., Springfield Mill, James Whatman Way, Maidstone, Kent ME14 2LE, UK.

1 Introduction

This chapter discusses some of the issues associated with the purification of proteins and related biomolecules from microbial sources. The emphasis in the examples described is on the scale-up of these processes to produce commercial quantities of the proteins. The protein purification steps can best be separated into upstream and downstream processes. For simplicity, we will delineate downstream processes into those using clarified feedstocks suitable for chromatography whereas upstream processes are those preceding clarification, i.e. cell harvesting, lysis, inclusion body concentration, etc. The scope of this chapter is restricted to downstream processes with upstream processes being well described elsewhere (1, 2).

2 Chromatographic processes

There are a range of chromatographic techniques available for purification of biomolecules from microbial sources including:

- salt precipitation
- ion exchange
- hydrophobic interaction
- affinity
- size exclusion
- chiral
- thiophilic interaction

Each of these techniques exploits the unique physicochemical properties of individual proteins and it is not intended to discuss the theory behind these techniques further as these have been adequately reported elsewhere (3, 4). Of these techniques those routinely scaled-up to process scale are ion exchange, hydrophobic interaction, size exclusion, and to a lesser degree affinity. Ion

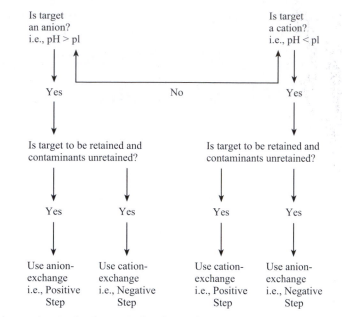

Figure 1 Approaches to development of an ion exchange chromatographic process.

exchange, hydrophobic interaction, and affinity chromatography are adsorptive techniques enabling selective adsorption/desorption of one or more components of the feedstock thereby effecting purification and/or concentration. From a practical standpoint there are several approaches to adsorption chromatography and a strategy for ion exchange is described in *Figure 1*. It is clear that dependent on the nature of the feedstock and the process requirements anion exchange or cation exchange chromatography may be carried out in positive or negative steps. Examples of each of these approaches are reported elsewhere (3). In a similar manner hydrophobic interaction chromatography or affinity chromatography may be carried out as either a positive or negative step, the strategy being dictated by the chromatographic objectives.

Each adsorptive stage, be it a positive or negative step, may be carried out in a column or batch contacting system. The features of each type of contacting system have been examined in the process scale ion exchange separation of hen egg white proteins (5), and some practical aspects of large scale column chromatography have been reported elsewhere (6, 7).

In any event key considerations in the selection of chromatographic technique and the mode of operation should be made.

2.1 Positive or negative steps

A positive adsorptive step gives opportunity not only for selective adsorption, i.e. purification but also concentration, if the elution volume is less than the feedstock volume. Where selective desorption is required a gradient, step or continuous, may be preferred. It should be noted that the product in its elution buffer, may be the feedstock for a subsequent downstream process. Therefore consideration as to whether its mobile phase is suitable for the next step needs to be made, otherwise extensive and costly additional operations such as diafiltration, desalting, solvent extraction, etc. may be necessary.

A negative step, whilst not providing concentration often gives significant purification and has the advantage that the mobile phase is consistent through-out the chromatographic stage. This may eliminate the need for mobile phase modifications to the eluate.

2.2 Batch or column contactors

For positive steps, especially those using gradient elution a column-based system would be recommended. However in negative steps a batch system may offer process benefits in terms of time and cost since there is no requirement for high resolution product desorption (5).

In the purification of biomolecules from microbial sources, these considera-tions discussed above require examination in order to optimize the process routing. Consideration should also be given to process validation, particularly when the product may be used for pharmaceutical application or other regu-lated processes. The subject of process validation is outside of the scope of this article, but aspects of chromatographic validation have been described elsewhere (8–10).

2.3 Gradient elution

In any adsorptive process, the target biomolecule adsorbs to a stationary phase and is then selectively desorbed, prior to elution from the matrix. In highly selective separations, for example affinity, the adsorptive interaction is gener-ally so selective that a simple step elution is acceptable. However, in the case of hydrophobic interaction chromatography or ion exchange chromatography many proteins may bind to the adsorbent during the adsorption stage. In order to effect selective desorption, it is necessary to carefully control the elution buffer conditions. One simple approach is to carry out gradient elution in a column contacting system. Linear gradients are the most widely used although convex or concave ones may be generated. A linear gradient involves a linear change in mobile phase composition from buffer A to buffer B as a function of time or volume. Linear gradients are simple to generate and we have reported their effectiveness at volumes up to 400 litres using both cellulose and agarose-based ion exchangers in 25 litre chromatography columns (11). A protocol for generating and running a linear gradient is given here.

Protocol 1

Generation and running of a linear gradient

Equipment and reagents

- Two tanks of equal dimensions
- Equal volumes of buffers A and B
- Agitator

Protocol 1 continued

Method

1 Stand both tanks on level ground adjacent to one another.

2 Fill one tank with buffer A and fill the other tank with an identical volume of buffer B.

3 Place the agitator into the tank containing buffer A.

4 Connect tanks A and B at the base.

5 Develop the gradient by pumping elution buffer from tank A.

6 As buffer A is withdrawn from tank A, it is replaced by buffer B creating a linear gradient which increases with respect to buffer B with time.

The principles associated with *Protocol 1* should work for ion exchange where buffers A and B may have differing ionic strengths or pH and for hydrophobic interaction chromatography where they may have different polarities.

3 Ion exchange protein purifications from microbial cell culture

In this section examples will be given for the ion exchange purification of several proteins from microbial sources. These are intended to be illustrative and support the underlying principles discussed above.

3.1 Isolation of yeast enzymes

Yeast is a rich source of proteins, particularly enzymes, which can be isolated for various applications. To be effective, cell debris needs to be removed prior to conventional packed bed chromatography. One approach which we have investigated for the release of glyceraldehyde 3-phosphate dehydrogenase (GAPDH), 3-phosphoglycerate kinase (PGK), and triosephosphate isomerase (TIM) is as follows.

Protocol 2

Release of GAPDH, PGK, and TIM from yeast cells

Equipment and reagents

- Dried baker's yeast cells (Sigma, DCL Yeast Ltd.)
- Grade 541 and GF/C filter paper (Whatman)
- 1.0 μm nitrocellulose membrane (Whatman)
- Sodium hydroxide
- Acetic acid
- Ammonium sulfate
- Storage buffer: 50 mM sodium phosphate buffer pH 7.0, 2 mM EDTA

Protocol 2 continued

Method

1   Disperse 100 g dried yeast in 1000 ml of deionized water and add sodium hydroxide until the pH is 10, to lyse the cells.

2   During cell lysis, maintain the pH at 10 by addition of sodium hydroxide. Lyse the cells for 3 h.

3   Adjust the pH to 5.2 using acetic acid.

4   Filter the suspension through a Whatman grade 541 filter. The filtrate is further clarified by filtration through a Whatman GF/C filter.

5   Add ammonium sulfate to the filtrate to give a 20% (w/v) solution and store at 4°C for 12 h.

6   Remove the precipitate by filtration through a Whatman GF/C filter.

7   Add ammonium sulfate to the filtrate to give a 47% (w/v) solution and store at 4°C for at least 12 h.

8   Collect the precipitate by filtration through a Whatman 1.0 μm nitrocellulose membrane.

9   Dissolve the precipitate in 100–200 ml of storage buffer. This can be stored at −20°C for further use.

Table 1 Enzyme profile during their release from yeast cells[a]

Stage of experiment	Total activity (units)		
	GAPDH	PGK	TIM
Lysate	463 050	284 850	2 025 000
20% $(NH_4)_2SO_4$ supernatant	302 460	285 420	860 250
47% $(NH_4)_2SO_4$ precipitate	189 255	208 310	752 950

[a] Based on 100 g baker's yeast cells.

A typical profile of the three enzymes during their release from the yeast cells is summarized in *Table 1*.

The yeast extract can then be processed using packed bed chromatography. In the case of ion exchange the extract requires a change in mobile phase composition, to remove the residual ammonium sulfate, thereby reducing the ionic strength to one more appropriate for protein adsorption.

The yeast extract may then be chromatographed by anion exchange chromatography to effect partial separation of the various enzymes. Dependent on the target enzyme and its required degree of purity, the chromatographer has scope to manipulate the elution buffer to optimize the desired separation. As an example TIM can be isolated by anion exchange chromatography as described.

Protocol 3

Preparation of yeast extract for ion exchange chromatography

Equipment and reagents

- Yeast extract (see *Protocol 2*)
- Loading buffer: 10 mM triethanolamine–HCl buffer pH 9, 2 mM EDTA
- Visking dialysis tubing (Sigma)

Method

1 Fill the Visking dialysis tube with the yeast extract.

2 Dialyse against 10 volumes of loading buffer at 4°C for 12 h.

3 Repeat step 2 a further two or three times until the conductivity and pH of the loading buffer, following dialysis is identical to its starting value.

4 Collect the dialysed yeast extract and use for subsequent chromatography.

Protocol 4

Isolation of TIM from yeast extract

Equipment and reagents

- Dialysed yeast extract (see *Protocol 3*)
- DE52, QA52, or Express-Ion Q (Whatman)
- Loading buffer (see *Protocol 3*)
- Wash buffer (same as loading buffer)
- Elution buffer: loading buffer containing 500 mM NaCl
- PGK and TIM assayed according to Sigma protocol

Method

1 Equilibrate either DE52, QA52, or Express-Ion Q in loading buffer to constant pH and conductivity. Adjust to a 20–30% (w/v) slurry.

2 Pack DE52, QA52, or Express-Ion Q in a 1.5 cm i.d. chromatography column to give a bed height of 10 cm.

3 Load 40 ml of dialysed yeast extract to the column of ion exchanger at a flow rate of 2–4 ml/min.

4 Wash the column with 100 ml of wash buffer at a flow rate of 2–4 ml/min.

5 Elute bound material using a linear gradient of 200 ml wash buffer and 200 ml elution buffer at a flow rate of 2–4 ml/min. Monitor absorbance of the eluate at 280 nm.

6 Collect fractions for subsequent assay.

Table 2 Purification of TIM from extract of baker's yeast using QA52

Stage of experiment	PGK activity (units)	TIM activity (units)	TIM recovery (%)
Loading	7080	37 120	–
Wash	7207	48	–
Elution	550	33 360	89.9

Following *Protocols 2–4* using QA52, we isolated TIM free of PGK activity. In this example PGK only weakly bound to QA52, and was released during the wash step. TIM bound to the QA52 and eluted during the salt gradient to give ~90% recovery. These data are summarized in *Table 2*.

3.2 Isolation of amylase from *Bacillus subtilis*

The enzyme α-amylase hydrolyses the α-1,4-linkages within the starch molecule to yield maltose, maltotriose, and α-dextrin. Starch hydrolysis is of importance in the food industry for a range of applications including sweeteners, osmotic pressure, thermostability, saccharification, etc. Amylase is commercially available from several microbial hosts including *Bacillus subtilis* var. *amyloliquefaciens*, *Bacillus lichenoformis*, *Aspergillus niger*, and *Rhizopus* spp. The properties of each of these forms of amylase particularly with regard to thermostability are determined by their microbial source (12).

We have looked into the ion exchange chromatography of α-amylase from 'Nervanase', a commercial extract of *B. subtilis* var. *amyloliquefaciens*. This procedure requires two stages, sample clarification and anion exchange chromatography. There are two approaches to sample clarification as described below.

Protocol 5

Clarification of Nervanase by ammonium sulfate precipitation

Equipment and reagents

- Nervanase (ABM/Rhone Poulenc)
- Equilibration buffer: 20 mM Tris–HCl buffer pH 8.7
- Ammonium sulfate
- Amylase was assayed using the Phadebas system (Pharmacia)

Method

1 Adjust Nervanase to 65% (w/v) with ammonium sulfate.

2 Collect precipitated material by centrifugation and dissolve in equilibration buffer.

3 Equilibrate the precipitated fraction with equilibration buffer by dialysis at 4 °C (see *Protocol 3*) for subsequent chromatography.

This clarification protocol gave almost complete recovery of amylase in a feedstock appropriate for column chromatography. However for industrial scale application this process was considered impractical due to the large usage of ammonium sulfate. For example, to prepare 250 litres of feedstock would require 162.5 kg of ammonium sulfate, creating a significant effluent disposal consideration.

In order to overcome this issue we identified a diafiltration protocol for feedstock clarification.

Protocol 6

Clarification of Nervanase by diafiltration

Equipment and reagents

- Nervanase (ABM/Rhone Poulenc)
- 10 000 MWCO UF membranes (Filtron Omega) in a Pellicon UF apparatus (Millipore)
- Equilibration buffer (see *Protocol 5*)
- Amylase assay (see *Protocol 5*)

Method

1 Diafilter 290 litres of Nervanase through a 10 000 MWCO membrane against water.

2 Diafilter Nervanase (from step 1) against equilibration buffer to equilibrium.

This clarification protocol gave almost complete recovery of amylase in a feedstock appropriate for large scale column chromatography. In order to partially purify the amylase, a study using 290 litres of diafiltered Nervanase containing 1.148 kg total protein and 574 000 units of amylase was carried out.

Protocol 7

Anion exchange chromatography of Nervanase

Equipment and reagents

- Diafiltered Nervanase (see *Protocol 6*)
- Equilibration buffer: 20 mM Tris–HCl pH 8.7
- Elution buffer 1: equilibration buffer containing 250 mM NaCl
- Elution buffer 2: equilibration buffer containing 500 mM NaCl
- Express-Ion Q (Whatman)
- Moduline 45 cm i.d. glass chromatography column (Millipore)

Method

1 Equilibrate ~ 25 kg Express-Ion Q in equilibration buffer to give constant pH and conductivity. Adjust to a 30% (w/v) slurry.

2 Pack the Express-Ion Q in the 45 cm i.d. column at 15 psi to give a 25 litre bed (16 cm bed height).

Protocol 7 continued

3 Load 290 litres of Nervanase feedstock onto the column at a flow rate of 150 cm/h. Wash the column with equilibration buffer until baseline absorbance at 280 nm is reached.

4 Elute bound amylase using 100 litres of elution buffer 1 at a flow rate of 150 cm/h.

5 Regenerate column using 75 litres of elution buffer 2 at a flow rate of 150 cm/h.

Following *Protocol 7*, using Express-Ion Q, we isolated a total of 369 g protein in the 0.25 M NaCl step, containing 408 320 units of amylase. This corresponds to a 71% recovery with 2.2-fold purification and 3-fold concentration. These data are summarized in *Table 3*. The amylase purified thus far could be further fractionated if necessary by other chromatographic steps.

Table 3 Purification of amylase from Nervanase using Express-Ion Q

Stage of experiment	Amylase activity (units)	Protein (g)	Amylase recovery (%)
Loading	574 000	1148	–
Wash	99 615	452	–
Elution 1	408 320	369	71
Elution 2	17 100	137	–

3.3 Isolation of DNA-modifying enzymes

In order for the molecular biologist to carry out recombinant DNA techniques, it is necessary to have available a range of DNA-modifying enzymes. These include restriction enzymes (endonucleases) which cleave DNA at defined nucleotide sequences, ligases to join nucleic acids together, and polymerases to replicate DNA molecules *in vitro*. The type II endonucleases have become essential tools for all recombinant DNA experiments. Type II restriction enzymes recognize and bind to discrete, rotationally symmetrical DNA sequences in double-stranded DNA and cleave at these sites. The only cofactor requirement is Mg^{2+} ions. These enzymes are found throughout the bacterial kingdom and it is necessary to purify each enzyme from a different bacterial species. They are named by taking the first letter from the genus name followed by the first two letters of the species name (13). For example, the enzyme *Eco*RI comes from *E. coli* strain RY13. The roman numeral I designates this enzyme as the first enzyme isolated from this strain. Other examples are the enzyme *Pst*I from *Providencia stuartii*, *Kpn*I from *Klebsiella pneumoniae*, and *Alu*I from *Arthrobacter luteus*.

DNA ligases are essential in recombinant DNA experiments as they catalyse the joining together of DNA fragments generated by type II restriction enzymes. The ligase most commonly used for this purpose is T_4 DNA ligase encoded by the *E. coli* bacteriophage T_4. This enzyme is usually prepared from *E. coli* strains carrying a cloned version of this gene. The T_4 DNA ligase has the ability to join together DNA fragments with both sticky and blunt ends, making T$_4$ DNA ligase a highly versatile enzyme in cloning strategies.

DNA polymerases have a wide variety of uses in molecular biology for labelling DNA *in vitro*, for DNA sequencing, and importantly for the amplification of small amounts of DNA *in vitro* by the polymerase chain reaction (PCR). Both mesophilic phage and bacterial enzymes are used, in addition to an increasing number of enzymes from thermophilic organisms. The plethora of DNA-modifying enzymes now required by molecular biologists necessitates the use of a generic procedure for purification which is relatively specific for DNA-binding proteins but rapid and capable of coping with many different bacterial extracts. It has been found that Whatman phosphocellulose (P11) fulfils the above criteria (14). This phosphocellulose is a cation exchanger based on fibrous cellulose that also acts as a pseudo-affinity medium for enzymes that bind to nucleic acids, e.g. endonucleases, kinases, and polymerases.

A protocol for single stage isolation of DNA-modifying enzyme from bacterial cells is described in *Protocol 8*.

Protocol 8

Isolation of DNA-modifying enzymes from bacterial cells

Equipment and reagents

- Bacterial cell cultures grown to log phase (see *Table 4* and ref. 14)

- Phosphocellulose (P11) (Whatman)

- Assay systems: restriction digests, ligation, polymerase assays (see ref. 14)

- Extraction buffer 1: 10 mM potassium phosphate buffer pH 7.5, 7 mM 2-mercaptoethanol

- Elution buffer: extraction buffer, 1 M salt

- Lysozyme (Sigma)

Method

1 Harvest cells from a 1 litre culture by centrifugation at 12 000 *g* for 10 min.

2 Resuspend cell pellet in 50 ml extraction buffer, at 4 °C.

3 Repellet cells by centrifugation at 12 000 *g* for 10 min.

4 Resuspend cell pellet in 20 ml extraction buffer.

5 Add 20 mg lysozyme and lyse cells by incubation at 37 °C for 10 min.

6 Chill the suspension on ice and sonicate for 30 sec, followed by 30 sec cooling. Repeat this cycle a further nine times.

7 Clarify the lysate by centrifugation at 50 000 *g* for 40 min at 4 °C.

8 Apply the supernatant to a 15 ml column of P11, previously equilibrated with extraction buffer at a flow rate of 30 cm/h.

9 Wash the column with extraction buffer until baseline absorbance at 280 nm is reached.

10 Elute bound material using a linear gradient of 100 ml extraction buffer and 100 ml of elution buffer, at a flow rate of 30 cm/h.

11 Assay eluate for enzyme activity.

Table 4 DNA-modifying enzymes prepared using P11 chromatography

Source of enzyme	Enzyme name
E. coli RY13	EcoRI
Haemophilus influenzae Rd	HindII, HindIII
Providencia stuartii	PstI
Serratia marcescens	SmaI
Bacillus amyloliquefaciens H	BamHI
B. globigii	BglI, BglII
Klebsiella pneumoniae	KpnI
Arthrobacter luteus	AluI
Streptomyces albus G	SalGI
S. achromogenes	SacI, SacII
Proteus vulgaris	PvuI, PvuII
Bacillus caldolyticus	BclI
Haemophilus aegyptius	HaeII, HaeIII
H. haemolyticus	HhaII
Haemophilus parainfluenzae	HpaI, HpaII
Staphylococcus aureus 3A	Sau3A
Streptomyces phaechromogenes	SphI
Thermus aquaticus	TaqI
Xanthomas badrii	XbaI
Xanthomonas holicola	XhoI, XhoII
E. coli carrying cloned T_4 DNA ligase gene	T_4 DNA ligase
E. coli carrying cloned DNA polymerase gene	DNA polymerase I

This protocol has been found to be effective for isolation of 22 different DNA-modifying enzymes summarized in *Table 4* (14). In this type of purification absolute protein homogeneity is not essential. What is crucial however is that the enzyme preparation is free of contaminating enzyme activity or inhibitors. Assuming this to be the case, the presence of other proteins may be of little significance.

4 Ion exchange nucleic acid purification from microbial culture

Nucleic acids provide the templates coding for the proteins associated with all cellular functions. DNA is a polymer of deoxyribonucleotides and RNA is a polymer of ribonucleotides. DNA and RNA are anions at neutral pH and can therefore be isolated by anion exchange chromatography. The DEAE–cellulose paper, Whatman DE81 has been reported to bind DNA fragments (15) and tRNA (16). Plasmid DNA can be isolated from a cell lysate using absorption chromatography to a DEAE–cellulose (17).

Purified DNA can be used for a variety of purposes including PCR, sequencing, and for the emerging opportunities in gene therapy where plasmids expressed in microbial hosts are required to be isolated in large quantities.

Plasmid DNA is routinely isolated from bacterial cell lysates and a protocol using an anion exchange agarose bead containing a paramagnetic component is described (18). In this example the plasmid pBluescript is used, but the protocol should have widespread suitability.

Protocol 9

Isolation of plasmid DNA from bacterial cells

Equipment and reagents

- DEAE–Magarose® (Whatman)
- Magnetic separator stand (Biometra)
- STET buffer: 10 mM Tris–HCl pH 8.0, 100 mM NaCl, 1 mM EDTA, 1% (w/v) Triton X-100
- Wash buffer: 10 mM Tris–HCl pH 8.0, 400 mM NaCl, 1 mM EDTA
- Elution buffer: 10 mM Tris–HCl pH 8.0, 1 M NaCl, 1 mM EDTA
- Co-precipitant buffer: 7.5 M ammonium acetate

- TE buffer: 10 mM Tris–HCl pH 8.0, 1 mM EDTA
- Absolute ethanol
- E. coli JM109 cells expressing pBluescript
- Resuspension buffer: 50 mM Tris–HCl pH 8.0, 10 mM EDTA, 400 μg/ml ribonuclease A
- Lysis buffer: 200 mM NaOH, 1% (w/v) sodium dodecyl sulfate
- Precipitation buffer: 3 M potassium acetate buffer pH 5.5

Method

1 Grow *E. coli* JM109 cells expressing the plasmid pBluescript to late log phase in Luria-Bertani broth containing 100 μg/ml ampicillin. Harvest bacterial cells from 1.5 ml of cell culture by centrifugation at 10 000 g for 1 min.

2 Resuspend the pellet in 100 μl resuspension buffer.

3 Lyse the cells by mixing 200 μl lysis buffer with the cell suspension. Place on ice for 5 min.

4 Precipitate genomic DNA and other contaminants by addition of 150 μl precipitation buffer, previously cooled to 4°C. Stand on ice for 5 min and centrifuge at 10 000 g for 5 min.

5 Add the supernatant to 150 μl of 4% (w/v) DEAE–Magarose in STET buffer. Mix for 5 min at room temperature.

6 Immobilize the beads in a magnetic separator stand and discard the supernatant.

7 Wash the beads by resuspension in 400 μl of wash buffer.

8 Repeat step 6.

9 Desorb bound DNA using 200 μl elution buffer.

10 Immobilize the beads and collect the supernatant.

Protocol 9 continued

11 Precipitate the DNA by addition of 500 µl absolute ethanol, pre-cooled to −20°C, and 20 µl co-precipitant buffer. Store at −20°C for 10 min.

12 Collect the pellet by centrifugation at 15 000 g for 30 min at 4°C, and discard the supernatant. Wash the pellet with 50 µl of 70% (v/v) cold ethanol and recentrifuge at 15 000 g for 10 min at 4°C.

13 Discard the supernatant and allow the pellet to air dry at room temperature for 30 min.

14 Redissolve the pellet in 50 µl TE buffer.

15 The pBluescript DNA is ready for subsequent use.

Microbial cell lysates are relatively viscous and may present problems in a column operation. On the other hand the paramagnetic component of a chromatographic support facilitates rapid immobilization in a batch chromatographic step. The Magarose beads described in *Protocol 9* respond very rapidly to a magnetic field, typically being immobilized from a suspension in less than 10 seconds, leaving a clear supernatant which can readily be removed by aspiration or decantation. This enables the chromatographic protocols to be carried out rapidly, with minimal mobile phase carry-over from step-to-step resulting in short process time, economic use of reagents, and most importantly a decreased likelihood of product contamination.

The DNA binding properties of DEAE–Magarose for pBluescript are summarized in *Table 5*. In a plasmid DNA isolation, following an alkaline lysis procedure, DEAE–Magarose (4% (w/v) suspension, 150 µl) isolated approximately 8.2 µg of pBluescript DNA from 1.5 ml *E. coli* JM109 cell culture. As indicated by A_{260}/A_{280} ratio the DNA was of high purity. Agarose gel electrophoresis confirmed this to be the case. Preliminary indications have shown that the sequence of plasmid DNA isolated using DEAE–Magarose is readable at over 900 bases. The DNA eluted from the DEAE–Magarose is readily collected by precipitation with cold ethanol, giving a visible pellet. In certain cases other ion exchange-based plasmid DNA isolation kits require the use of isopropanol precipitation, which although effective does give a less clearly visible DNA pellet which may result in yield losses due to practical difficulties in handling this pellet.

Table 5 Binding of pBluescript DNA to DEAE–Magarose

Volume of beads used (µl)	DNA yield (µl)	A_{260}/A_{280} ratio
150	8.2 ($n = 5$)	1.94

References

1. Mackay, D. (1996). In *Downstream processing of natural products. A practical handbook* (ed. M. S. Verrall), p. 11. John Wiley & Sons, Chichester.
2. Keshavarz-Moore, E. (1996). In *Downstream processing of natural products. A practical handbook* (ed. M. S. Verrall), p. 41. John Wiley & Sons, Chichester.

3. Levison, P. R. (1991). In *Process-scale liquid chromatography* (ed G. Subramanian), p. 131. VCH, Weinheim.

4. Lowe, C. R. (1979). *An introduction to affinity chromatography*, p. 276. Elsevier Biomedical Press, Amsterdam.

5. Levison, P. R. (1993). In *Preparative and production scale chromatography* (ed. G. Ganetsos and P. E. Barker). p. 617. Marcel Dekker, New York.

6. Levison, P. R., Badger, S. E., Toome, D. W., Butts, E. T., Koscielny, M. L., and Lane, L. (1991). In *Upstream and downstream processing in biotechnology III* (ed. A. Huyghebaert and E. Vandamme), p. 3.21. The Royal Flemish Society of Engineers, Antwerp.

7. Levison, P. R. (1996). In *Downstream processing of natural products. A practical handbook* (ed. M. S. Verrall), p. 179. John Wiley & Sons, Chichester.

8. Sofer, G. K. and Nystrom, L.-E. (1991). *Process chromatography, a guide to validation*, p. 1. Academic Press, London.

9. Levison, P. R., Badger, S. E., Jones, R. M. H., Toome, D. W., Streater, M., Pathirana, N. D., *et al.* (1995). *J. Chromatogr. A*, **702**, 59.

10. Levison, P. R., Streater, M., Jones, R. M. H., and Pathirana, N. D. (1996). In *Validation practices for biotechnology products* (ed. J. K. Shillenn), p. 44. ASTM, West Conshohocken.

11. Levison, P. R., Jones, R. M. H., Toome, D. W., Badger, S. E., Streater, M., and Pathirana, N. D. (1996). *J. Chromatogr. A*, **734**, 137.

12. Best, D. J. (1985). In *Biotechnology. Principles and applications* (ed. I. J. Higgins, D. J. Best, and J. Jones), p. 143. Blackwell Scientific Publications, Oxford.

13. Smith, H. O. and Nathans, D. (1973). *J. Mol. Biol.*, **81**, 419.

14. Ward, J. M., Wallace, L. J., Cowan, D., Shadbolt, P., and Levison, P. R. (1991). *Anal. Chim. Acta*, **249**, 195.

15. Kaczosowski, T., Sektas, M., and Furmanek, B. (1993). *Biotechniques*, **14**, 900.

16. Harris, C. L. and Kolanko, C. J. (1989). *Anal. Biochem.*, **176**, 57.

17. Van Huynh, N., Motte, J. C., Pilette, J. F., Decleire, M., and Colson, C. (1993). *Anal. Biochem.*, **211**, 61.

18. Levison, P. R., Badger, S. E., Dennis, J., Hathi, P., Davies, M. J., Bruce, I. J., *et al.* (1998). *J. Chromatogr. A*, **816**, 107.

Chapter 6
Protein purification from milk

Richard Burr
Food Science, New Zealand Dairy Research Institute, Private Bag 11 029, Palmerston North, New Zealand.

1 Introduction

This chapter illustrates examples of techniques used in the purification of common milk proteins. It outlines some of the specific factors to consider in the handling, purification, and analysis of milk components. Simple protocols for the purification of major bovine milk proteins are provided and some newer developments that are applicable to the isolation of milk proteins are discussed.

2 Milk proteins

Milk is a complex mixture of proteins, lipids, carbohydrates, vitamins, and minerals structured to provide a complete diet for infant mammals (*Table 1*). Despite the range of compositional differences found in the milk of various species (1), it is useful to use bovine milk to illustrate the salient features that are common to most milks.

The bovine mammary gland produces six major secretory protein products: α_{S1}-casein (α_{S1}-CN), α_{S2}-casein (α_{S2}-CN), β-casein (β-CN), κ-casein (κ-CN), α-lactalbumin (α-La), and β-lactoglobulin (β-LG) and a host of minor protein products. These include serum albumin, immunoglobulins, and several enzymes, the more abundant being lactoferrin, lactoperoxidase, and xanthine oxidase. Genetic polymorphisms of each of the major milk proteins may also be observed. During milk synthesis two alleles are expressed for each protein. Differences in

Table 1 Interspecies comparison of milk composition (approximate % by weight)

Component	Human	Bovine	Ovine	Caprine
Water	88.0	86.6	81.3	87.7
Protein	0.27 (casein) 0.73 (whey)	2.6 (casein) 0.6 (whey)	5.5	2.9
Fat	3.8	4.1	7.4	4.5
Carbohydrate	7.0	5.0	4.8	4.1
Minerals and salts	0.2	0.7	1.0	0.8

the genetic code for each allele give rise to genetic variants of each milk protein. Milk proteins are also subject to post-translational modifications (e.g. glycosylation and phosphorylation) and intra- and intermolecular interactions. The presence of blood plasmin and plasminogen in milk results in a number of small (proteose-peptones) and large (γ-casein) peptides as a result of post-translational proteolysis.

2.1 Micelle structure

In milk, caseins are aggregated into large colloidal micelles. Bovine milk contains 30–36 g/litre of total protein of which about 80% is integrated into casein micelles in the approximate molar ratio of 3 : 3 : 1 : 1 for α_{S1}-CN : β-CN : α_{S2}-CN : κ-CN. About 5% (w/v) of the micelle is composed of calcium, phosphate, and other small ions. Casein macromolecular assemblies are built up out of many individual molecules and form a colloidal solution. A common feature of casein proteins is the esterification of phosphoric acid to a hydroxyl group. The phosphoric acid binds calcium and magnesium and some complex salts to form bonds between and within molecules. The structure of casein micelles appears to rely on the hydrophobic interaction between casein protein species stabilized by electrostatic bridges between charged areas of the casein subunits, and calcium and phosphate bound primarily to phosphoserine residues.

Calcium salts of α_{S1}-CN, α_{S2}-CN, and β-CN are near insoluble in water, while those of κ-CN are readily soluble. κ-CN occupies a key surface position in the micelle with the negatively charged macropeptide terminus projecting into the solvent as a flexible 'tail'. This provides the micelle with steric and electrostatic stabilization effects. If the hydrophilic N-terminal is cleaved from κ-CN (e.g. by chymosin) the solubility of the micelle decreases, the proteins aggregate and casein curd is formed.

2.2 Major milk protein groups

Three main groups of proteins in milk are distinguished by their differing properties.

(a) The caseins are easily precipitated from milk by acidification.

(b) Serum proteins remain soluble during acid casein precipitation.

(c) Milk fat globular membrane proteins are associated with the lipid fraction.

The major bovine milk proteins traditionally have been subdivided into casein or whey proteins on the basis of their solubility at pH 4.6 at 20°C. Precipitated material under these conditions is classified as casein; the remaining soluble proteins are classified as whey proteins. The advent of starch gels, and later polyacrylamide gel electrophoresis (PAGE) allowed the isolation and identification of individual families of milk proteins. With the determination of their primary sequence it was then possible to define them on the basis of their chemical structure, rather than solubility (*Figure 1*). It is now recognized that some material that remains soluble during casein precipitation (proteose peptones) is

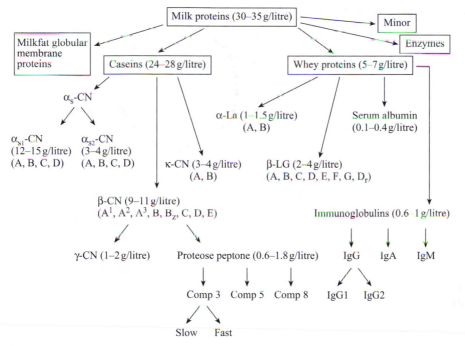

^aGentetic varients of milk proteins in brackets

Figure 1 Distribution of major milk proteins and peptides in bovine milk.

derived from proteolysis of casein proteins. Whey proteins include α-La, β-LG, serum albumin (SA), immunoglobulins (Igs), and a host of minor proteins and enzymes. Another class of proteins are associated with the milk fat globular membrane and include a range of enzymes and lipoproteins.

The key physical–chemical characteristics of the major bovine milk proteins are listed in *Table 2*.

2.2.1 Caseins

Caseins are the major (75–80%) proteins of all ruminant milks and occur in all mammalian milks. Bovine casein concentration is between 2.5% and 3.2% (w/v) in milk. Caseins have a high proportion of charged residues clustered at one end of the molecule. This gives them a marked amphipathic nature and is thought to be important in maintaining stability of the micelle. The phosphoserine residues in caseins bind calcium strongly. Addition of 0.25 M Ca^{2+} at pH 7 and 37°C results in precipitation of the 'calcium sensitive' caseins—those with a high proportion of phosphoserine residues (α_{S1}-CN, α_{S2}-CN, and β-CN). The soluble 'calcium insensitive' fraction is predominantly κ-CN (2).

i. α_{S1}-CN

α_{S1}-CN is a polypeptide of 199 residues in bovine species. The exception is the A variant (rarely seen in *Bos taurus* milk) which differs by a deletion of residues

Table 2 Biochemical characteristics of bovine milk proteins

Protein	M_r^a	pHi[b]	Conc. (g/litre)	ε^{280} (M^{-1} cm^{-1})	RM[j]
α-s1 CN A	22069	4.16–4.40	12–15	11.3[c]	1.22
α-s1 CN B-8P	23614	4.23–4.47		10.5,[c] 10.1[d–g]	1.13
α-s1 CN C	23543	4.27–4.49	10.6[c]		1.10
α-s2 CN A-11P	25229	4.83–5.13	3–4	14.0,[f] 11.1,[c] 11.0[c]	1.00
β-CN A1–5P	24023	4.68–4.96	9–11	4.5,[f] 4.6[c–f]	0.65, 0.69
β-CN A2–5P	23983	4.60–4.84			0.65, 0.69
β-CN A3–5P	23974	4.50–4.74			0.65, 0.69
β-CN B-5P	24092	4.78–5.10			0.61, 0.65
κ-CN A-1P	19036	5.43–5.81	3–4	9.5,[c] 10.5,[f] 12.2[d,e,g]	0.45, 0.40
κ-CN B1-P	19004	5.54–6.12			0.40, 0.36
β-LG A	18363	4.64–4.90	2–4	9.5,[d–g] 9.7[d]	
β-LG B	18277	4.72–4.98		10.0[d]	
β-LG C	18286	4.77–5.13			
α-La B	14146	4.66–4.89	0.6–1.7	20.1,[e] 20.9[d–g]	
BSA	66267		0.2–0.4	6.14,[d] 6.2,[d] 6.3,[d] 6.49,[d] 6.6,[d,f,g] 6.61,[d] 6.8[d,e]	
Ig's			0.5–1.8	13.7,[d] 14.0,[i] 12.1,[h] 13.5[h]	

[a] Calculated from primary amino acid sequences published by Eigel *et al.* (93).
[b] Observed isoionic points from Siebert *et al.* (85).
[c] Fox (94).
[d] Fasman (95).
[e] Chemical Rubber Company (96).
[f] Walstra and Jenness (97).
[g] McKenzie (98).
[h] Fox (99).
[i] Sigma Chemical Company (100).
[j] Relative mobility to the migration of α$_{S2}$-CN A-11P in alkaline PAGE. From Swaisgood (101).

14–26. Three hydrophobic regions are discernible. Additionally the region 41–80 contains a cluster of phosphoseryl residues. These observations suggest a dipolar structure with a globular hydrophobic domain at one end, and a highly solvated and charged domain at the other. Amongst the caseins there is a strong sequence homology in these hydrophobic and phosphoseryl regions. This conservation of structural domains suggests that these proteins may have evolved from a common ancestral gene.

ii. α_{S2}-CN

α_{S2}-CN consists of 207 residues, possibly existing as dimers linked via a disulfide bond (3). 11 serine residues are phosphorylated. A cluster of negative charges at the N-terminus and positive charges at the C-terminus create a strong dipolar arrangement, suggesting that electrostatic interactions may be an important factor in influencing structural characteristics.

iii. β-CN

The β-CN consists of 209 residues, with a highly charged domain clearly separated from a hydrophobic domain. As with the α_S-CNs a cluster of phosphoseryl residues occurs near the N-terminus. The unusually high frequency of prolyl residues most likely influences the number of β-turns present (4, 5).

iv. κ-CN

In comparison with the other caseins κ-CN is notable for its carbohydrate moieties attached via threonyl residues and the absence of phosphate clusters. Consequently, κ-CN does not bind Ca^{2+} to the same degree as other caseins; thus its solubility is relatively independent of this ion. The amphipathic nature of this milk protein has been appreciated for many years due to the specific chymosin-catalysed hydrolysis of the Phe105–Met106 bond. This releases the polar glycomacropeptide from κ-CN, destabilizing the casein micelle, and results in clotting of milk. Approximately two-thirds of κ-CN molecules are glycosylated. Considerable heterogeneity is seen in this protein, with at least 17 forms varying in carbohydrate content and type identified. κ-CN, as isolated from milk, occurs in the form of a mixture of polymers linked by intermolecular disulfide bonding. There is evidence to suggest that β-LG is associated with κ-CN in the micelle via disulfide linkages.

2.2.2 Proteose peptones

Proteose peptones encompass a range of glycoproteins and phosphoproteins derived from proteolysis of milk proteins. The predominant proteose peptones are of β-CN origin.

2.2.3 Whey proteins

Whey protein is a term often used as a synonym for milk-serum proteins. The major families of proteins included in this class are β-LG, α-La, serum albumin, and immunoglobulins. Proteose peptones (soluble at pH 4.6) were included in this group prior to their identification as being a proteolysis product of other milk proteins.

i. α-La

α-La is a polypeptide consisting of 123 amino acids, incorporating four intramolecular disulfide bridges. Minor forms of α-La have been reported, most containing some form of carbohydrate moiety (hexosamine, mannose, galactose, fucose, N-acetylglucosamine, N-acetylgalactosamine, and N-acetylneuramininc acid). α-La forms part of the enzyme complex galactosyltransferase in the synthesis of lactose.

ii. β-LG

β-LG is the principal whey protein in bovine milk. It consists of 162 residues, of which five are cysteine. Four of these are involved in intramolecular disulfide bonds and one as a free thiol. In the pH range from 5.7 to 7.0 all genetic variants have been shown to primarily exist as dimers. Below pH 3.5 β-LG is predominantly monomeric. Dissociation also occurs above pH 7.5 suggesting that carboxyl groups are involved in the process (6). β-LG is known to bind a range of non-polar ligands such as retinol and fatty acids.

iii. Serum albumins

SA accounts for approximately 1.2% of the total milk protein present. SA prepared from milk is physically and immunologically identical to blood SA. Analysis by isoelectric focusing (IEF) shows considerable microheterogeneity, although no variant species are known to exist. Bovine SA consists of 582 amino acid residues with 17 intramolecular disulfide bonds and a single free thiol.

iv. Immunoglobulins

The immunoglobulins form an extremely heterogeneous family of proteins and are classified primarily by immunochemical criteria. Four classes have been identified in bovine milk (IgG, IgA, IgM, and IgE) all existing as glycoprotein monomers or polymers of a basic unit composed of four polypeptide chains linked covalently by disulfide bonds.

2.2.4 Milk fat globule membrane proteins

Milk fat exists as small globules or droplets dispersed in the milk serum. The emulsion is stabilized by a thin membrane 5–10 nm thick which surrounds the globule. The membrane is composed of phospholipids, lipoproteins, cerebrosides, proteins, nucleic acids, enzymes, trace elements, and water. Fat globule membrane proteins adhere to the surface of the fat globules and are released upon aggregation of the fat globules, e.g. when cream is churned to butter.

2.2.5 Enzymes

A diverse range of enzymes may be found in milk. They may be located in several partitions:

- free in solution
- associated with or part of membrane fractions
- associated with casein micelles
- as part of microsomal particles

The major categories are oxidoreductases, transferases, hydrolases, lyases, isomerases, and ligases. For further information the reader is referred to more detailed reviews (6–9).

2.2.6 Hormones and bioactive peptides

As with enzymes, an array of hormones and biologically active peptides may also be found in milk. Milk secreted in the first few days after parturition contains the highest concentration of hormones and growth factors. Accurate quantitative determinations of many of these substances are difficult as they often interact with interfering substances. Detailed reviews have been published by Campana and Baumrucker (10), and Koldovský and Štrbák (11).

3 Factors to consider in preparation and handling milk

Milk proteins are subject to a number of factors during purification that may adversely affect the final product. The relative importance of each factor is determined by the end use of the target protein. Hence the design of an effective purification strategy must preserve the essential characteristics of the protein.

3.1 Source

Environmental and genetic influences contribute to the variations seen in milk composition and volume produced. During the course of lactation the composition of milk alters considerably, particularly in regard to immunoglobulin content in early lactation, and enzyme activity in late lactation. The end use of the protein(s) and quantities required will dictate the appropriate species, phenotype, and stage of lactation.

3.2 Proteolysis

Raw milk contains not only endogenous enzymes, but a range of micro-organisms that hydrolyse milk proteins. Proteolysis is reduced by pasteurization and/or cooling milk to 2–4°C. Note that psychotrophic micro-organisms still reproduce at 4°C and significant breakdown of fats and proteins may occur over a short period of time. Purification should generally begin with fresh milk. If milk is to be stored for any length of time prior to processing it is advisable to add a protease inhibitor such as 15 mM phenylmethylsulfonyl fluoride (PMSF) or 15 mM ε-amino-hexanoic acid and refrigerate at 0–4°C.

3.3 Solubility

Differential solubility is frequently used in milk protein purifications. In chromatographic separations the solubility of a sample during purification is of paramount importance to avoid column fouling. Caseins generally require chaotropic agents such as urea or guanidine hydrochloride to disrupt the strong protein–protein interactions in the casein micelle. Often a final step of purification may involve removal of salts, buffer exchange, or pH adjustment prior to lyophilization or frozen storage. For whey proteins this may be easily accomplished chromatographically on a column of Sephadex G25. Casein solutions have a tendency to precipitate at high protein concentrations, low ionic strength, and acidic pH values. If there is any doubt as to the solubility of a casein solution during desalting it is safer to dialyse.

3.4 Protein modification

Thermal or chemical denaturation of milk proteins are frequently used to precipitate milk proteins. Although attractively simple the methods may result in
some irreversible modification of proteins. After weak heat or chemical denaturation a protein may regain the semblance of its original structure. However the fidelity of the original structure is questionable. Alterations of the pH or
ionic environment will alter the hydration state and zeta potential of a protein.
Maillard products characterize the modification of reactive amines that have
been heated in the presence of carbohydrates. Heating of milks is also known to
induce the interaction of β-LG with κ-CN via sulfydryl-disulfide interchange.
Extremes of pH and heat may also induce racemization, dephosphorylation, or
hydrolysis (particularly of κ-CN).

3.4.1 Urea purification

Chaotropic agents are frequently used in the purification of milk proteins to
disrupt protein–protein interactions characteristic of caseins. Dissociation is
frequently achieved with concentrated solutions of urea or guanidine hydrochloride. Urea is attractive due to its lower cost, particularly when purifying
large quantities of casein proteins. Urea spontaneously forms cyanate and
ammonia particularly at elevated temperatures and pH values (*Figure 2*). Cyanate
readily carbamylates reactive amines (lysine side chains, N-terminal amines, and
cysteine). For purification of caseins, whether by selective precipitation or by
chromatographic means, it is useful to prepare a volume of purified stock 8 M
urea, diluting and/or adding buffers as required. Cyanate may be removed from
urea by passing through a mixed bed ion exchange matrix such as Duolite MB
6113 (BDH Chemicals). Alternatively separate columns of anion exchanger (e.g.
Duolite A113) and cation exchanger (e.g. Duolite C225) may be used in series
with the advantage of simplifying the regeneration of exhausted resins. Further

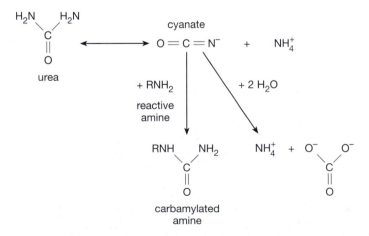

Figure 2 Cyanate formation and carbamylation of reactive amines.

cyanate formation is inhibited by reducing the pH of purified stock urea solution by addition of glacial acetic acid to pH 5.0 and storage at 4°C.

3.5 Drying and storage

Lyophilization (freeze-drying) is often regarded as the best means to preserve proteins. However there is some debate about the effects of lyophilization on protein conformation. Whey proteins in general are reasonably stable in mild acid solutions. In some cases it may be preferable to freeze solutions for short-term storage.

Whey proteins desalt and lyophilize readily. The pure proteins resolubilize well even at elevated concentrations. At higher concentrations casein proteins tend to precipitate upon dialysis and are sometimes more difficult than whey proteins to dissolve after lyophilization. Casein solutions may still contain an appreciable amount of urea even after extensive dialysis against pure water. Solubility problems with κ-CN in particular may be overcome by dialysis against a weak, volatile buffer (e.g. dilute ammonium acetate or ammonium bicarbonate) prior to lyophilization.

4 Fractionation of milk proteins

Often two or more methods of protein purification are employed to obtain substantially pure material. Milk is frequently separated into casein and whey fractions by isoelectric precipitation prior to higher resolution techniques. Classical methods of milk protein separations are based on selective solubility differences in various solvent systems. The protocols outlined (3 and 4) yield substantially purified individual proteins using simple apparatus and commonly available solvents. The development of physically stable, high resolution chromatographic matrices has resulted in a number of methods published yielding highly purified forms of individual proteins from casein or whey feed stock. Ultrafiltration, phase separation, and preparative electrophoretic techniques have expanded the range of fractionation approaches available to the biochemist. A summary of techniques and selected references are included in *Table 3*.

4.1 Fractionation of casein and whey

Historically casein curd has been precipitated from milk as a result of the enzymatic cleavage of the chymosin sensitive bond in κ-CN. Release of the stabilizing hydrophilic terminus of κ-CN from the surface of the micelle culminates in the aggregation of micelles into casein curd—the preliminary step to making cheese. Acidification is more commonly used to separate casein and whey proteins for purification, avoiding the additional problems of removing any added enzymes.

Distinct changes occur in the casein micelle upon acidification. Between pH 6.6 and 5.5, casein micelles are homogeneously distributed in solution. Below pH 5.5 calcium phosphate is released, reducing its binding role to the extent

Table 3 Summary of methods used to purify milk proteins

Method	Notes	Refs
Differential solubility	Substantial purification possible, cheap, requires minimum of equipment. Suited to initial separation prior to higher resolution techniques.	2, 12–15
Chromatographic separations		
(a) Size exclusion	Limited resolution of caseins—all similar in mass. Can utilize associations—κ-CN in dissociating, non-reducing conditions predominantly polymeric, β-LG dimeric at pH 5.7–7.0. Capacity limited by sample volume. Good desalting, buffer exchange step.	17–19
(b) Ion exchange	Baseline resolution of individual proteins possible. High capacity, rapid. Suited for both preparative and analytical separations. Fractions may require desalting.	21–26
(c) Reverse-phase	High resolution, fast. Solvents may limit use of proteins. Limited life span of column when separating caseins?	27, 28
(d) Hydrophobic interaction	Limited resolution.	29
(e) Hydroxyapatite	Specific for phosphate containing proteins. Limited life span of column?	30, 31
(f) IMAC	Potentially useful for whey protein separations. Application for caseins unknown.	32–36
(g) Thiophilic	Binds free thiol containing proteins—β-LG; κ-CN, and α_{S2}-CN under reducing conditions.	37, 38
(h) Dye binding	Mechanism poorly understood.	39
(i) Immobilized ligand	Limited by specificity of ligand. Potential for β-LG purification.	40
(j) Chromatofocusing	High resolution, expensive or complex buffer system, may require desalting.	41
Heat, pH, or solvent	Cheap, minimum of equipment. Potential modifications to proteins.	42, 43
Ultrafiltration	Limited by similarity of milk protein masses. Can utilize low temperature solubility of β-CN, non-aggregation of β-LG under mild heat to isolate.	44
Electrophoretic behaviour		
(a) Preparative IEF	Potentially high resolution, contamination with carrier ampholytes?	
(b) Preparative PAGE	High resolution, may require buffer change/desalting.	
Ultracentrifugation	Very gentle, little disturbance to protein environment. Limited by volume of rotor, centrifuge times, resolves casein micelles from whey.	45
Aqueous phase separation	Limited resolution.	46, 101

that β-CN and κ-CN are substantially released from the micelle. At about pH 5.2 the micellar frameworks of α_{S1}-CN and α_{S2}-CN form extensive fields of aggregated structures. From pH 5.2 to 4.8 the aggregation stage is followed by a consolidation phase. Finally between pH 4.8 and 4.5 rearrangement and aggregation of these particles occur as casein curd.

4.1.1 Defatting

Lipids are often difficult to deal with effectively in aqueous systems. Effective defatting of milk is usually accomplished by centrifugation. It may also be achieved by solvent extraction. However, this exposes the proteins to potentially denaturing solvents and is infrequently used. Skimming of milk may be accomplished in a simple separator. Even small scale separators are capable of processing multi-litre volumes of skim milk within minutes. Further reduction of fat content may be achieved by a second pass through the separator, altering the skimmer cone parameters and/or centrifugation.

Protocol 1

Preparation of skim milk

Equipment and reagents

- Fresh whole milk
- Vacuum source
- Separator (e.g. Elecrem 80, Vertrieb & Kundendienst) or centrifuge

Method

1   Warm fresh whole milk to 50°C.

2   (a) Process larger sample volumes through a separator following the manufacturer's instructions as to adjustment of separator plates. A more complete separation of cream from whole milk may be achieved by passing a second time through the separator and/or reducing the flow rate.

   (b) Centrifuge smaller volumes (or as a second step for larger volumes) at 5000 g for 15 min (unrefrigerated).

3   Remove the supernatant fat layer. This is easily accomplished using a pipette tip connected to a vacuum source. If the sample is allowed to cool the fat layer will harden, making this step more difficult.

4.1.2 Casein precipitation

Protocol 2

Casein precipitation

Equipment and reagents

- Skim milk (*Protocol 1*)
- 1 M HCl
- 1 M NaOH
- Cheese cloth
- Centrifuge

Method

1   Adjust the temperature of skim milk to 20°C. Adjust the pH to 4.6 by slow addition of 1 M HCl[a] with continuous stirring.[b]

2   Stand at room temperature for 30 min. Separate the precipitated casein curd from the whey supernatant by filtering through several layers of cheese cloth.

3   Wash the casein curd twice with cold, deionized water and finally with warm deionized water, crumbling the curd into small particles to remove residual whey trapped in the curd mass.

4   Adjust the whey from step 2 to pH 7.0 by addition of 1 M NaOH. Centrifuge at 5000 g for 15 min to pellet precipitated calcium and phosphate complexes, and fine particles of casein. Decant off clarified whey.

[a] May be substituted by 1 M H_2SO_4.

[b] Milk that has been exposed to enzyme activity may precipitate poorly or not at all.

Wet acid casein may be stored frozen at $-20\,°C$ for short periods of time. Casein may be re-solubilized in water at pH 7.0 by slow addition of 1 M NaOH. It may be desirable to dialyse re-solubilized casein curd against a volatile buffer, e.g. 10 mM ammonium acetate pH 7.0 prior to lyophilization for longer-term storage. Whole casein may also be prepared by salt precipitation with either Na_2SO_4 (47) or $(NH_4)_2SO_4$ (48).

4.1.3 Ultracentrifugation

If the native conformation of a milk protein or preservation of the casein micelle is of prime importance then ultracentrifugation of milk at $60\,000$–$80\,000\,g$ for 60 min at 20–$25\,°C$ is a means of separating casein micelles from whey without subjecting the milk to non-physiological ranges of temperature, pH, solvents, or ionic concentrations. The translucent whey may be separated from the casein pellet and the fat layer by carefully puncturing the centrifuge tube with a needle and syringe, and withdrawing the whey. The casein pellet and fat layer may then be separated by cutting the centrifuge tube between the two.

Protocol 3

Fractionation of casein proteins by differential solubility

Equipment and reagents

- Acid casein (*Protocol 2*)
- 3.5 M H_2SO_4
- Purified 8 M urea stock solution[a]
- Centrifuge
- $(NH_4)_2SO_4$
- 2 M NaOH

- 1 M ammonium acetate in 75% ethanol
- Ethanol
- Dialysis tubing
- Dialysis buffer:10 mM ammonium acetate pH 7.0

Method

The sequential dilution of protein–urea solutions requires advance calculations of volumes and the appropriate sized vessels.

Protocol 3 continued

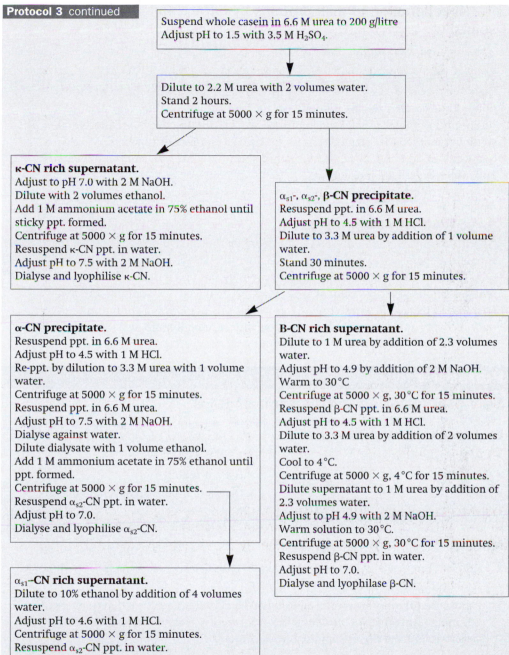

Suspend whole casein in 6.6 M urea to 200 g/litre
Adjust pH to 1.5 with 3.5 M H_2SO_4.

Dilute to 2.2 M urea with 2 volumes water.
Stand 2 hours.
Centrifuge at 5000 × g for 15 minutes.

κ-CN rich supernatant.
Adjust to pH 7.0 with 2 M NaOH.
Dilute with 2 volumes ethanol.
Add 1 M ammonium acetate in 75% ethanol until
sticky ppt. formed.
Centrifuge at 5000 × g for 15 minutes.
Resuspend κ-CN ppt. in water.
Adjust pH to 7.5 with 2 M NaOH.
Dialyse and lyophilise κ-CN.

α_{s1}-, α_{s2}-, β-CN precipitate.
Resuspend ppt. in 6.6 M urea.
Adjust pH to 4.5 with 1 M HCl.
Dilute to 3.3 M urea by addition of 1 volume
water.
Stand 30 minutes.
Centrifuge at 5000 × g for 15 minutes.

α-CN precipitate.
Resuspend ppt. in 6.6 M urea.
Adjust pH to 4.5 with 1 M HCl.
Re-ppt. by dilution to 3.3 M urea with 1 volume
water.
Centrifuge at 5000 × g for 15 minutes.
Resuspend ppt. in 6.6 M urea.
Adjust pH to 7.5 with 2 M NaOH.
Dialyse against water.
Dilute dialysate with 1 volume ethanol.
Add 1 M ammonium acetate in 75% ethanol until
ppt. formed.
Centrifuge at 5000 × g for 15 minutes.
Resuspend α_{s2}-CN ppt. in water.
Adjust pH to 7.0.
Dialyse and lyophilise α_{s2}-CN.

B-CN rich supernatant.
Dilute to 1 M urea by addition of 2.3 volumes
water.
Adjust pH to 4.9 by addition of 2 M NaOH.
Warm to 30°C
Centrifuge at 5000 × g, 30°C for 15 minutes.
Resuspend β-CN ppt. in 6.6 M urea.
Adjust pH to 4.5 with 1 M HCl.
Dilute to 3.3 M urea by addition of 2 volumes
water.
Cool to 4°C.
Centrifuge at 5000 × g, 4°C for 15 minutes.
Dilute supernatant to 1 M urea by addition of
2.3 volumes water.
Adjust to pH 4.9 with 2 M NaOH.
Warm solution to 30°C.
Centrifuge at 5000 × g, 30°C for 15 minutes.
Resuspend β-CN ppt. in water.
Adjust pH to 7.0.
Dialyse and lyophilase β-CN.

α_{s1}–**CN rich supernatant.**
Dilute to 10% ethanol by addition of 4 volumes
water.
Adjust pH to 4.6 with 1 M HCl.
Centrifuge at 5000 × g for 15 minutes.
Resuspend α_{s2}-CN ppt. in water.
Dialyse and lyophilise α_{s2}-CN.

[a] See notes Section 3.4.1 for urea purification.

[b] Samples prepared in urea may contain significant quantities of urea even after extensive
dialysis. Urea may be removed by desalting on a Sephadex G25 or similar column. See notes
Section 3.3 on casein solubility.

4.2 Fractionation of milk proteins by differential solubility

4.2.1 Casein proteins

Several differential solubility fractionation methods have been developed for the preparation of casein proteins. Whilst a general method for the purification of all of the casein proteins is described, more selective and direct purifications of specific proteins have been published (2, 13, 15, 16, 49, 50). Differential solubility methods are attractive as no specialized equipment is necessary, large quantities of protein may be processed in a short-time interval, and the solvents and salts required are inexpensive.

4.2.2 Whey proteins

As with caseins, several differential solubility methods have been developed to isolate the major whey proteins (51–54). Of particular use is the purification of β-LG to 90%+ purity from acid whey (*Protocol 4*) by salt precipitation of non β-LG whey proteins. Precipitation of α-La may be achieved by addition of 0.5% trichloroacetic acid to whey at pH 7.1. However this method is a compromise of poor recovery at < 0.5% and contamination with serum albumin at > 0.5% (53).

Protocol 4

Salt precipitation purification of β-LG

Equipment and reagents

- Clarified whey (*Protocol 2*)
- Centrifuge
- NaCl
- 3.5 M HCl
- 3.5 M NaOH
- Dialysis tubing
- Dialysis buffer: 10 mM ammonium acetate pH 6.0

Method

1   Add NaCl in small aliquots to 7 g total per 100 ml whey, allowing NaCl to dissolve completely before adding the next aliquot.

2   Adjust the pH to 2.0 by careful addition of 3.5 M HCl. Stand at room temperature for 20 min. Centrifuge at 10 000 g for 15 min to isolate the β-LG rich soluble supernatant from the precipitated whey proteins.

3   Resuspend the non-β-LG whey protein precipitate in water. Dialyse and lyophilize.

4   Add 23 g of NaCl per litre of decanted supernatant to give a final concentration of 30% (w/v). Readjust the pH to 2.0 by careful addition of 3.5 M NaOH. Stand at room temperature for 20 min. Centrifuge at 10 000 g for 15 min. Resuspend the β-LG precipitate in water and dialyse before lyophilization.

5 Chromatography of milk proteins

Chromatographic methods have been widely used for the isolation of milk proteins. Several reviews have been published (6, 17, 55–57) that may be used as a guide to select an appropriate system (*Table 3*). High purity milk proteins may be prepared by combining an initial precipitation (*Protocols 3* and *4*) with a high resolution chromatographic step (*Protocols 5* and *6*).

5.1 Ion exchange

Diethyaminoethyl (DEAE)–cellulose and DEAE–Sepharose have been used in the past for the purification of both casein (58, 59) and whey proteins (48, 60). Casein is normally treated with 2-mercaptoethanol to reduce disulfide bonds in κ- and α_{S2}-CN, and urea included in the elution buffer to dissociate the individual caseins. Pure preparation of β- and α_{S1}-CN may be obtained by careful pooling of selected fractions. κ-CN tends to elute in a broad band, reflecting the various glycosylated forms. α_{S2}-CN is difficult to isolate with any degree of purity from α_{S1}-CN (*Figure 3*).

Cellulose-based media tend to compress and swell during elution and regeneration, often requiring the column to be repacked at frequent intervals. They have been largely superseded by more rigid support material allowing higher flow rates and improved resolution (23, 61–63). These methods were partially successful in resolving α_{S1}- and α_{S2}-CN on a milligram scale. Ng-Kwai-Hang and Dong Chin (24) developed a method utilizing a PROTEIN PAK DEAE-15 HR column which adequately resolved all the caseins from 250 mg of whole casein.

Mono-Q has been widely used for small scale purification and analysis of whey proteins (21, 22, 61, 64–66) (*Figure 4*). Recovery of proteins is quantitative, reproducible, and fast. Consequently this method has been frequently adopted for routine analysis of whey proteins.

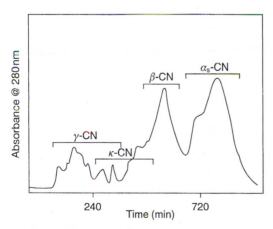

Figure 3 Elution profile of casein on Whatman DE-52 ion exchange column. Casein dissolved in 4.5 M urea, 10 mM imidazole pH 7.0 and reduced with 0.1% mercaptoethanol. Chromatography on a 2.6 × 70 cm column with a 0–0.4 M NaCl gradient in 4.5 M urea, 10 mM imidazole, 0.1% mercaptoethanol pH 7.0. Elution of each species indicated by bars.

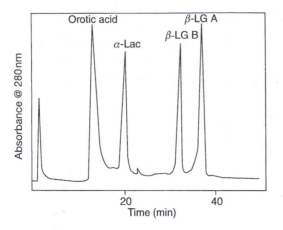

Figure 4 Elution profile of acid whey on a Pharmacia Mono Q HR 5/5 column, 20 mM piperazine buffer pH 6.0 with a 0–0.4 M NaCl gradient over 50 min. Figure adapted with permission from Humphrey and Newsome (22).

Cation exchange chromatography, using Spherosil S, has been used industrially to absorb whey proteins from acidic whey (67). Resolution of individual proteins is poor but large quantities of whey proteins may be isolated from non-protein components in a process stream at relatively low cost.

Preparative scale purification of gram quantities of casein proteins may be achieved using S-Sepharose Fast Flow media (*Protocol 5*).

Protocol 5

Cation exchange chromatography of casein proteins

Equipment and reagents

- Acid casein (*Protocol 2*)
- Buffer A: 6 M urea, 20 mM sodium acetate pH 5.0[a]
- Buffer B: 6 M urea, 20 mM sodium acetate pH 5.0, 1 M NaCl
- 2 M NaOH
- 2 M HCl

- Dithiothreitol
- S-Sepharose Fast Flow (Pharmacia)
- Binary gradient chromatography system
- Dialysis tubing
- Dialysis buffer: 10 mM ammonium acetate pH 7.0

Method

1 Dissolve acid casein in buffer A to 2% (w/v),[b,c] adjusting the pH to 7.0 by careful addition of 2 M NaOH.

2 Add 0.1% (w/v) DTT, stirring continuously overnight at room temperature.

3 Carefully titrate back to pH 5.0 with 2 M HCl. It is important not to exceed the minimum requirement of acid or base to meet the required pH of the solution.

Protocol 5 continued

Stand for 1 h at room temperature before centrifuging at 5000 g for 10 min, discarding any pelleted material.

4   Load 1/5 of the column volume (CV) of sample (e.g. 300 ml of sample for a 1500 ml column) at a flow rate of 1 CV/h.

5   Elute with a 3 CV gradient from 0% to 40% buffer B.

6   Wash the column with 0.5 CV of 100% buffer B, followed by equilibrating with 1 CV of 100% buffer A.

7   Collect eluate fractions as appropriate (see *Figure 5*).

8   Dialyse fractions against several changes of dialysis buffer and lyophilize.

[a] See notes (Section 3.4.1) on urea purification.

[b] Assume wet casein is approximately 50% protein by weight.

[c] Hollar *et al.* (23) recommends a concentration of 0.3–0.75%. We have found that concentrations of up to 5% are suitable. Above 5% the solution becomes viscous and some caseins may selectively precipitate.

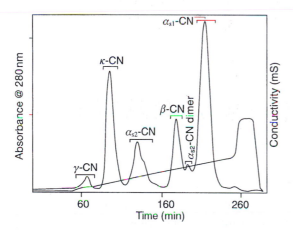

Figure 5 Ion exchange chromatography of whole casein on Pharmacia S-Sepharose Fast Flow. 5 g of acid precipitated casein dissolved in 6 M urea, 20 mM sodium acetate pH 5.0, reduced with 0.1% dithiothreitol at pH 7.0. Chromatography on a 15 ×113 cm column at 25 ml/min, 0–0.4 M NaCl gradient in 6.0 M urea, 20 mM sodium acetate pH 5.0 buffer. Elution of casein species indicated by bars.

5.2 Size exclusion chromatography

Size exclusion chromatography of milk proteins has been extensively investigated. Strange *et al.* (56) reviewed the more recent contributions to the literature. In general the resolution of individual caseins is poor due to their similarities in size. Casein micelles have been successfully isolated from other milk proteins using controlled pore glass beads (68–71) and Sephacryl S-1000 (72, 73). Some precipitation and subsequent column fouling may occur when using simulated

103

milk ultrafiltrate (SMUF) buffer as an eluent at near neutral pH values (6.7–7.0). This may be avoided by lowering the pH of the eluent to less than 6.7.

The major whey proteins may be satisfactorily resolved under neutral conditions (Protocol 6). At pH 7.0, β-LG exists predominantly as dimers. The size of the dimer (~ 36000 Da) is sufficiently different from α-La (~ 14000 Da) and serum albumin (~ 66000 Da) under these conditions to allow near baseline separation of the major components (*Figure 6*). Sephacryl S-200 and S-300 have been used for high purity isolations of xanthine oxidase, lactoferrin (74), lactoperoxidase (74), and immunoglobulins (75) from acid whey. Salt precipitation preparations of β-LG and other whey proteins from acid whey (*Protocol 4*) are frequently contaminated by other whey proteins. Size exclusion chromatography following salt precipitation purification simultaneously achieves both a high degree of purity and reduces salt content. The extremely reproducible nature of size exclusion chromatography is ideally suited to automation of the process. By combining these two methods, large quantities of purified whey proteins may be prepared with a minimum of labour.

Protocol 6

Size exclusion chromatography of whey proteins

Equipment and reagents

- Whey (*Protocol 2*), crude β-LG, or other whey proteins (*Protocol 4*)
- Pharmacia Superdex 75 26/60 Prep column
- Isocratic chromatography system
- Elution buffer: 20 mM phosphate, 30 mM NaCl pH 6.0
- Dialysis tubing
- Dialysis buffer: 10 mM ammonium acetate pH 6.0

Method

1 Dissolve crude β-LG or whey proteins in elution buffer up to 50 mg/ml. Adjust the whey from Protocol 2 to pH 6.0 with 1 M NaOH.

2 Centrifuge at 5000 *g* for 15 min, discarding any insoluble material.

3 Load up to 5 ml onto the column at a flow rate of 2.5 ml/min.

4 Elute at 2.5 ml/min, collecting appropriate fractions (see *Figure 6*).

5 Dialyse fractions against dialysis buffer and lyophilize.

5.3 Affinity chromatography

Several chromatographic methods have been reported in the literature that utilize affinity characteristics of milk proteins. The affinity of phosphate groups, found in α_{S1}-, α_{S2}-, β-, and to a lesser degree κ-CN, for calcium is utilized with hydroxyapatite resin, eluting with an increasing concentration of phosphate buffer (31, 76). Hydrophobic chromatography has been used to isolate whey proteins (29, 77) and caseins (78). Thiophilic adsorbents such as activated thiol

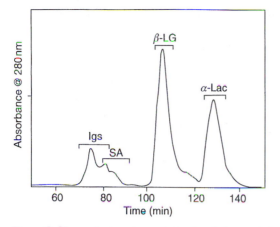

Figure 6 Size exclusion chromatography of whey proteins on Pharmacia Superdex 75 prep 2.6 × 60 cm column; 2.5 ml of a 10 mg/ml solution of whey applied and eluted at 2.5 ml/min with 20 mM phosphate, 30 mM NaCl pH 6.0 buffer. Elution of whey species indicated by bars.

Sepharose 4B may be utilized to selectively bind proteins that contain free thiols (reduced κ-CN, α_{S2}-CN, β-LG, serum albumin), eluting with cysteine (37). Immobilized metal affinity chromatography (IMAC) has been used to isolate whey proteins (36, 79), enzymes (80, 81), and immunoglobulins (34).

A number of specific affinity methods have been developed for isolation of β-LG, enzymes, and immunoglobulins. Detailed discussion of the diverse range of ligands is beyond the scope of this chapter.

5.4 Reverse-phase chromatography

Reverse-phase high performance liquid chromatography (RP-HPLC) has been extensively utilized for analytical separations of both casein and whey proteins. A range of large pore (300 Å) columns have proven to be useful, from C_2 to C_{18}, resolution generally improving with increasing hydrocarbon chain length. Solvent systems are usually water–acetonitrile–trifluoroacetic acid, employing an acetonitrile gradient to elute bound material. This technique is particularly useful for analysis of genetic variants of milk proteins as many of the casein (31) and whey protein variants (82) are well-resolved.

Reverse-phase is less attractive as a large scale preparative technique. Excellent resolution is offset by the likelihood of some irreversible denaturation of proteins. Significant changes in protein conformation, reductions in enzymatic and immunological activity have been reported following reverse-phase separation (83). The cost, handling, and disposal of solvents on a larger scale must be considered.

5.5 Chromatofocusing

Pearce and Shanley (41) achieved good resolution of serum albumin, β-LG variants A and B, and α-La from acid whey on a column of Pharmacia Polybuffer 94 media with a pH gradient of 5.2–4.2. The sensitivity of the method is reflected in the narrow range of isoelectric points of the compounds separated (Table 2). The cost of the complex Polybuffer 74, used in the creation of the pH gradient, and the requirement for subsequent desalting limits the applicability of chromato-focusing for large scale separations.

6 Identification and analysis of purity

Historically the identification and purity of milk protein fractions has been assessed by alkaline urea gel electrophoresis, predominantly in polyacrylamide gels. Developments in high resolution, analytical ion exchange, and reverse-phase matrices have added an array of chromatographic techniques that are attractive due to their speed of analysis, reproducibility, and the ability to accurately quantitate peak areas. Whitney (84) and Swaisgood (102) provide concise reviews of chromatographic separations of milk proteins and caseins respectively. Gonzales-Liano *et al.* (57) published a more extensive review of analytical HPLC and FPLC techniques for the analysis of dairy products.

6.1 Polyacrylamide gel electrophoresis

Several polyacrylamide gel electrophoresis PAGE methods are used in the analysis of milk proteins as no single method resolves all milk proteins satisfactorily. Three gel systems are commonly used for the analysis of milk protein purity. Examples of each may be seen in *Figure 7*.

(a) Alkaline urea PAGE resolves casein proteins well, including the majority of casein protein polymorphisms. A high concentration of urea is incorporated into both the gel system and sample buffer. The chaotropic nature of urea disrupts the association between casein species, allowing proteins to migrate as monomers. However the inclusion of urea tends to result in diffuse bands for whey proteins.

(b) Sodium dodecyl sulfate (SDS)–PAGE is a good general type of gel for milk protein analysis, characterized by tight, well-defined bands and low background staining. SDS is a negatively charged surfactant. It is used to disrupt non-covalent bonds through its ability to absorb to hydrophobic and positively charged sites on proteins. Most proteins bind similar amounts of SDS on a mass basis. The SDS–protein complexes may then be separated on the basis of molecular size of the protein. Caseins are tightly clustered within a small region due to their similarity in mass. Samples are normally reduced by treatment with 2-mercaptoethanol to disrupt intra- and intermolecular disulfide bonds. When calculating the mass of proteins from SDS–PAGE, anomalous behaviour is observed in the relative mobilities of some caseins.

(c) Non-dissociating, non-reducing (or 'native') PAGE resolves proteins on the basis of charge, size, and conformation. Whey proteins are well resolved but caseins appear as a broad smear. Background staining is somewhat higher than alkaline urea or SDS–PAGE.

Protocol 7

PAGE analysis of milk proteins

The PAGE methods described below are based on discontinuous buffer systems, utilizing a stacking gel to concentrate proteins into a narrow zone prior to entering the resolving gel. Three systems are described. Refer to discussion above for selection of the appropriate system for the sample to be analysed.

Gel methods are optimized for $60 \times 100 \times 0.75$ mm gels.

Equipment and reagents

- Vertical slab gel electrophoresis equipment (e.g. Bio-Rad Mini-Protean II system)
- Stock acrylamide: 30% (w/v) acrylamide, N,N′-methylene-bis-acrylamide (BIS) (37.5:1), electrophoresis grade[a]
- 10% (w/v) ammonium persulfate in water
- Tetramethylenediamine (TEMED)
- 0.4% (w/v) bromophenol blue in water
- Glycerol
- 2-Mercaptoethanol
- 10% (w/v) SDS in water
- Stain: 0.05% Coomassie Brilliant Blue R in isopropanol/acetic acid/water (2.5:1:6.5, by vol.)
- Destain: isopropanol/acetic acid/water (1:1:8, by vol.)
- Resolving gel buffers
  (a) Alkaline urea: 0.38 M Tris–HCl pH 8.8, 4.5 M urea (adjust with 1 M HCl)
  (b) SDS: 1.5 M Tris–HCl pH 8.8 (adjust with 6 M HCl)
  (c) Native: 3 M Tris–HCl pH 8.8 (adjust with 6 M HCl)

- Stacking gel buffers
  (a) Alkaline urea: 90 mM Tris–HCl pH 8.4, 0.6 M urea, 90 mM boric acid, 2.5 mM EDTA
  (b) SDS: 0.5 M Tris–HCl pH 6.8 (adjust with 6 M HCl)
  (c) Native: 0.5 M Tris–HCl pH 6.8 (adjust with 6 M HCl)
- Sample buffers
  (a) Alkaline urea: 90 ml alkaline urea stacking gel buffer, 2 ml 0.4% bromophenol blue, 8 ml glycerol
  (b) SDS: 12.5 ml SDS stacking gel buffer, 50 ml water, 2.5 ml 0.4% bromophenol blue, 10 ml glycerol, 20 ml 10% SDS
  (c) Native: 20 ml native stacking gel buffer, 60 ml water, 2 ml 0.4% bromophenol blue, 8 ml glycerol
- Electrode buffers
  (a) Alkaline urea: 17.6 mM Tris–HCl pH 8.4, 17.8 mM boric acid, 0.55 mM EDTA
  (b) SDS: 25 mM Tris–HCl pH 8.6, 190 mM glycine, 3.5 mM SDS
  (c) Native: 25 mM Tris–HCl pH 8.3, 0.2 M glycine

A Resolving gel

Gel system	Alkaline urea	SDS	Native
Acrylamide stock	4.00 ml	5.30 ml	5.00 ml
Resolving buffer	5.95 ml	2.50 ml	1.25 ml
Water	3.75 ml	2.00 ml	–
10% SDS stock	–	100 µl	–
APS	50 µl	50 µl	50 µl
TEMED	5.0 µl	5.0 µl	5.0 µl

Protocol 7 continued

1 Measure the quantities described in the above table, less SDS, APS, and TEMED, for the required gel system into a 100 ml Buchner flask and degas under vacuum, with stirring for 15 min.

2 Add the SDS, APS, and TEMED gently swirling after each addition.

3 Deposit 3.30 ml of resolving gel solution in between the assembled glass plates of the gel equipment. Carefully overlay with 200 μl of water, pipetting down the sides of the glass plates.

4 Allow to polymerize at room temperature for at least 30 min before draining off the water.

B Stacking gel

Gel system	Alkaline urea	SDS	Native
Acrylamide	1.30 ml	0.65 ml	0.625 ml
Stacking buffer	8.65 ml	1.25 ml	1.25 ml
Water	–	3.05 ml	3.15 ml
10% SDS stock	–	50 ml	–
APS	50 μl	25 μl	25 μl
TEMED	10 μl	5 μl	5 μl

1 Measure the quantities described in the above table, less SDS, APS, and TEMED for the required gel system into a 100 ml Buchner flask and degas under vacuum, with stirring for 15 min.

2 Add the SDS, APS, and TEMED gently swirling after each addition.

3 Pipette stacking gel solution carefully on top of the polymerized resolving gel until the cavity is full.

4 Insert well-forming combs into the stacking gel solution, avoiding trapping air bubbles underneath the combs.

5 Allow to polymerize at room temperature for at least 1 h. Remove combs and overlay sample wells with enough water to fill completely.

6 Place inside a sealed plastic bag to prevent drying and shrinking of the cast gel and allow completion of polymerization overnight.

C Sample preparation

Sample	Alkaline urea	SDS	Native
Casein	0.6 mg/ml	0.6 mg/ml	
Liquid whey		Dilute 1:1	Dilute 1:1
Skim milk	Dilute 1:40	Dilute 1:40	Dilute 1:20
Purified proteins	0.1 mg/ml	0.1 mg/ml	0.1 mg/ml

1 Dilute samples in appropriate sample buffer according to the table above. Reduce alkaline urea and SDS samples with 20 μl mercaptoethanol/ml. Stand at room temperature for 1 h. Heat SDS samples in a boiling water-bath for 4 min.

2 Load 10 μl of sample per well.

Protocol 7 continued

D Electrophoresis conditions

Gel system	Voltage (V)[b]	Current (mA)[b]	Power (W)[b]	Time (h)[c]
Alkaline urea	210	70	6.5	1.4/1.7
SDS	210	70	6.5	0.9/1.1
Native	210	70	6.5	1.3/1.5

1 Run gel(s) using the appropriate conditions according to the conditions above.

2 Stain gel(s) for 1 h followed by two changes of destain. Refer to *Figure 7* for identification of proteins bands.

[a] Inconsistent results may occur with old acrylamide solutions. Polymerize and discard after one month.

[b] Limit values.

[c] Run times for one or two gels.

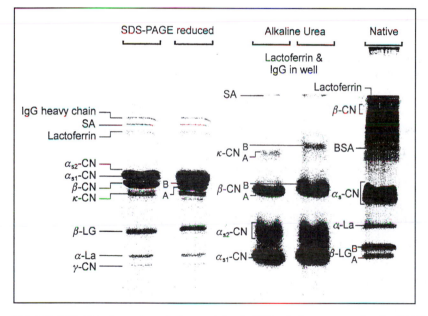

Figure 7 Migration of casein and whey proteins in SDS, alkaline urea, and native PAGE gel systems.

6.2 Isoelectric focusing

IEF methods have been developed that resolve all the major milk proteins and most polymorphisms (85–87). In these systems pre-cast IEF gels were modified by incubation in urea and ampholytes before use. Most protein species are resolved including variously phosphorylated and glycosylated forms, leading to a number of bands associated with each protein. The use of highly purified standards are recommended to distinguish between multiple bands attributed to a single protein and contaminating material.

6.3 Capillary electrophoresis

Capillary electrophoresis utilizes the principles of slab gel electrophoresis to resolve proteins in small diameter capillaries held in an electric field. Separation and quantitation can be performed in one step as in HPLC utilizing very small sample volumes. De Jong *et al.* (88) demonstrated good analytical separation of all major milk proteins at low pH in coated capillaries using a polymeric buffer additive. Other methods for analysis of whey proteins (89) and caseins (90) resolve most variant forms of each protein.

6.4 Analytical chromatography

Analytical chromatography of milk proteins is discussed in Section 5. Ion exchange (22, 61, 62) and reverse-phase chromatography (31, 82) are widely utilized due to their resolution, short analysis times, and reproducibility. Identification of protein species is normally based on a comparison of the elution time of protein standards and sample peaks.

6.5 Mass spectrometry

Recent advances in ionization technologies has resulted in the proliferation of relatively low cost mass spectrometers capable of protein mass analysis to an accuracy of 0.02% or better. The speed, precision, and compatibility of mass spectrometry with many separation techniques has resulted in many research laboratories routinely using mass spectrometry for identification and purity analysis of proteins (91). If the sequence of a protein is known, a comparison of observed mass and theoretical mass is often sufficient for identification.

The highest degree of certainty in identification of a protein is amino acid sequencing. Traditionally this has involved protein hydrolysis, separation of the peptide fragments, sequential N-terminal Edman degradation, and chromatographic identification of each released residue. The combination of tandem mass spectrometry (MS-MS) and advances in data analysis software has speeded up the process of primary sequence determination considerably.

On-line liquid chromatography-mass spectrometry (LC-MS) has the potential to determine the mass and purity of proteins during the process of chromatographic purification.

7 Developments

7.1 Preparative electrophoresis

The resolution of PAGE and IEF has been utilized in the development of small scale (mg to g) preparative electrophoretic systems. The Rotofor (Bio-Rad) IEF system uses either carrier ampholytes or soluble ionic compounds to establish a pH gradient in a series of chambers separated by 10 μm pore size screens. The IsoPrime (Hoefer) system avoids the additional step of removing ampholytes or buffers by immobilizing them in polyacrylamide membranes that separate the sample chambers.

Bio-Rad also manufactures preparative electrophoresis cells. Sample proteins applied onto cylindrical or tubular PAGE gels. During electrophoresis proteins focus into tight bands. Bands are continuously eluted from the gel into elution buffer and pumped to a fraction collector. PAGE methods may be transferred directly from vertical slab gel systems with modification of electrophoresis times.

Little information is available on preparative electrophoresis of milk proteins. The methods appear attractive for small scale purifications. Careful consideration should be given to elution solvent composition and protein concentrations to maintain sample solubility particularly with IEF. The inclusion of SDS in SDS–PAGE may require proteins to be further purified

7.2 Microfiltration and ultrafiltration

Progress in membrane technology has resulted in several fractionation methods of milk proteins being developed. In general filtration of milk proteins could be termed an enrichment process as opposed to a high resolution purification step.

Microfiltration and ultrafiltration has been applied in industrial applications to concentrate, demineralize, defat, and enrich protein containing products. By manipulation of filtration conditions, selective isolation of one or more components is possible. At low pH with moderate heat treatment (55°C, 30 min), α-La polymerizes and in the process traps other whey proteins with the exception of β-LG. The soluble β-LG may then be separated by 0.2 μm pore size microfiltration. The α-La may be isolated from the retentate by resolubilization at neutral pH followed by ultrafiltration with a 50 kDa cut-off membrane.

Microfiltration of skim milk concentrates micellar casein. Filtrate from a 0.2 μm pore size membrane resembles the composition of sweet whey without caseinomacropeptide, phospholipoproteins, or bacteria. At low temperatures (4°C) β-CN partially solubilizes from the casein micelle and may be separated by a 0.2 μm pore size membrane.

Affinity chromatography and microfiltration have been combined to separate lactoferrin and immunoglobulins from cheese whey. Heparin Sepharose, protein G Sepharose and protein G-bearing streptococcal cells were used as adsorbents to form affinity complexes with target proteins. These could be retained by 0.2 μm microfiltration membranes and separated from unbound whey proteins (92).

A concise overview of milk protein fractionation by micro- and ultrafiltration is provided by Maubois and Ollivier (44).

7.3 Aqueous two-phase partitioning

Aqueous two-phase partitioning using poly(ethylene glycol) (PEG)-sulfate and PEG-phosphate biphasic systems for ovine milk protein purification has been evaluated by Ortin *et al.* (46). Manipulating the PEG molecular weight, type of salt, pH, and salt concentration altered the partitioning of proteins. The partitioning effect was greatest for hydrophobic proteins, particularly at pH values close to the isoelectric point of the protein. A counter-current distribution of whey in the presence of the hydrophobic ligand palmitate-PEG and Ca^{2+} allowed substantial purification of β-LG in a single extraction. The mild separation conditions, low costs, and simplicity of this method are attractive for larger scale and continuous applications prior to higher resolution techniques.

Acknowledgements

I would like to acknowledge my colleagues at the NZDRI for useful discussion of aspects of this chapter during its preparation.

References

1. Ofterdal, O. T. and Iverson, S. J. (1995). In *Handbook of milk composition* (ed. R. G. Jensen), p. 749. Academic Press, San Diego.
2. Waugh, D. F. and Von Hipple, P. H. (1956). *J. Am. Chem. Soc.*, **78**, 4576.
3. Hoagland, P. D., Thompson, M. P., and Kalan, E. B. (1971). *J. Dairy Sci.*, **54**, 1103.
4. Andrews, A. L., Atkinson, D., Evans, M. T. A., Finer, E. G., Green, J. P., Phillips, M. C., *et al.* (1979). *Biopolymers*, **18**, 1105.
5. Graham, E. R. B., McLean, D. M., and Zviedrans, P. (1984). *Proc. Conf. Austr. Assoc. An. Breed. Genet., Adelaide*, **4**, 136.
6. Swaisgood, H. E. (1993). In *Advanced dairy chemistry, Vol. I. Proteins* (ed. P. F. Fox), p. 63. Elsevier Science Publishers, Essex.
7. Shahani, K. M., Harper, W. J., Jensen, R. G., Parry, R. M., and Zittle, C. A. (1973). *J. Dairy Sci.*, **56**, 531.
8. Kitchen, B. J. (1985). In *Developments in dairy chemistry. Vol 3. Lactose and minor constituents* (ed. P. F. Fox), p. 239. Applied Science Publishers, London.
9. Farkye, N. Y. (1992). In *Advanced dairy chemistry. 1. Proteins* (ed. P. F. Fox), p. 339. Elsevier Applied Science, London.
10. Campana, W. M. and Baumrucker, C. R. (1995). In *Handbook of milk composition* (ed. R. G. Jensen), p. 476. Academic Press, San Diego.
11. Koldovský, O. and Štrbák, V. (1995). In *Handbook of milk composition* (ed. R. G. Jensen), p. 428. Academic Press, San Diego.
12. Christensen, T. M. I. E. and Munksgaard, L. (1989). *Milchwissenschaft*, **44**, 480.
13. Fox, P. F. and Guiney, J. (1972). *J. Dairy Res.*, **39**, 49.
14. Brignon, G., Ribadeau-Dumas, B., and Mercier, J.-C. (1976). *FEBS Lett.*, **71**, 111.
15. Hipp, N. J., Groves, M. L., Custer, J. H., and McMeekin, T. L. (1952). *J. Dairy Sci.*, **35**, 272.
16. Zittle, C. A. and Custer, J. H. (1963). *J. Dairy Sci.*, **46**, 1183.
17. Yaguchi, M. and Rose, D. (1971). *J. Dairy Sci.*, **54** (12), 1725.
18. Dimenna, G. P. and Segal, H. J. (1981). *J. Liq. Chromatogr.*, **4**, 639.
19. Gupta, B. B. (1983). *J. Chromatogr.*, **282**, 463.
20. Bican, P. and Blanc, B. (1982). *Milchwissenschaft*, **41** (11), 700.
21. Hill, A. R., Manji, B., Kakuda, Y., Myers, C., and Irvine, D. M. (1987). *Milchwissenschaft*, **42** (11), 693.
22. Humphrey, R. S. and Newsome, L. J. (1984). *N. Z. J. Dairy Sci. Technol.*, **19**, 197.

23. Hollar, C. M., Law, A. J. R., Dalgleish, D. G., and Brown, R. J. (1991). *J. Dairy Sci.*, **74**, 2403.

24. Ng-Kwai-Hang, K. F. and Dong Chin. (1994). *Intl. Dairy J.*, **4**, 99.

25. Davies, D. T. and Law, A. J. R. (1977). *J. Dairy Res.*, **44**, 213.

26. Outinen, M., Tossavainen, O., and Syvaoja, E. L. (1996). *Lebensmittel Wissenschaft Technologie*, **29** (4), 340.

27. Carles, C. (1986). *J. Dairy Res.*, **53**, 35.

28. Bican, P. and Sphani, A. (1991). *J. High Res. Chromatogr.*, **14**, 287.

29. Chaplin, L. C. (1986). *J. Chromatogr.*, **363**, 329.

30. Donnelly, W. J. (1977). *J. Dairy Res.*, **44**, 612.

31. Visser, S., Slangen, K. J., and Rollema, H. S. (1986). *Milchwissenschaft*, **41**, 559.

32. Al-Mashikihi, S. A., Li-Chan, E., and Nakai, S. (1988). *J. Dairy Sci.*, **71**, 1747.

33. Scanff, P., Yvon, M., and Pelissier, J. P. (1991). *J. Chromatogr.*, **539**, 425.

34. Fukomoto, L. R., Li-Chan, E., Kwan, L., and Nakai, S. (1994). *Food Res. Intl.*, **27** (4), 335.

35. Kim, Y. J. and Cramer, S. M. (1991). *J. Chromatogr.*, **549**, 89.

36. Reid, T. S. and Stancavage, A. M. (1989). *BioChromatography*, **4** (5), 262.

37. Dall'olio, S., Davoli, R., and Russo, V. (1990). *J. Dairy Sci.*, **73**, 1707.

38. Nijuhuis, H. and Klostermeyer, H. (1975). *Milchwissenschaft*, **30**, 530.

39. Shimazaki, K. and Nishio, N. (1991). *J. Dairy Sci.*, **74**, 404.

40. Ena, J. M., Castillo, H., Sanchez, L., and Calvo, M. (1990). *J. Chrom. Biomed. Applicat.*, **525** (2), 442.

41. Pearce, R. J. and Shanley, R. M. (1981). *Austr. J. Dairy Tech.*, **36**, 110.

42. Manji, B. and Kakuda, Y. (1986). *Can. Inst. Food Sci. Technol. J.*, **19**, 163.

43. Murphy, B. F. and Mulvihill, D. M. (1988). *J. Soc. Dairy Technol.*, **41**, 22.

44. Maubois, J.-L. and Ollivier, G. (1991). In *New applications of membrane processes, IDF Special Issue 9201*, p. 15. International Dairy Federation, Brussels.

45. Waugh, D. F., Ludwig, M. L., Gillespie, J. M., Melton., B., Foley, M., and Kleiner, E. S. (1962). *J. Am. Chem. Soc.*, **84**, 4929.

46. Ortin, A., Muiño-Blanco, M. T., Calvo, M., Lopez-Perez, M. J., and Cebrian-Perez, J. A. (1992). *J. Dairy Sci.*, **75**, 711.

47. Wake, R. G. and Baldwin, R. L. (1961). *Biochim. Biophys. Acta*, **47**, 225.

48. McKenzie, H. A. (1971). In *Milk proteins, Vol. II* (ed. H. A. McKenzie), p. 258. Academic Press, New York.

49. Swaisgood, H. E. and Brunner, J. R. (1962). *J. Dairy Sci.*, **45**, 1.

50. McKenzie, H. A. and Wake, R. G. (1961). *Aust. J. Chem.*, **12**, 734.

51. Aschaffenburg, R. and Drewry, J. (1957). *Biochem. J.*, **65**, 273.

52. Armstrong, J. McD., McKenzie, H. A., and Sawyer, W. H. (1967). *Biochim. Biophys. Acta*, **147**, 60.

53. Fox, K. K., Holsinger, V. H., Posati, L. P., and Pallansch, M. (1967). *J. Dairy Sci.*, **50**, 1363.

54. Aschaffenburg, R. (1968). *J. Dairy Sci.*, **51**, 1295.

55. Ng-Kwai-Hang, K. F. and Grosclaude, F. (1992). In *Applied dairy chemistry 1: Proteins* (ed. P. F. Fox), p. 405. Elsevier Applied Science, London.

56. Strange, E. D., Malin, E. L., Van Hekken, D. L., and Basch, J. J. (1992). *J. Chromatogr.*, **624**, 81.

57. Gonzalez-Llano, D., Polo, C., and Ramos, M. (1990). *Lait*, **70**, 255.

58. Ribadeau-Dumas, B. (1961). *Biochim. Biophys. Acta*, **54**, 400.

59. Ribadeau-Dumas, B., Maubois, J. L., Mocquot, G., and Garnier, J. (1964). *Biochim. Biophys. Acta*, **82**, 494.

60. Armstrong, J. McD., Hopper, K. E., McKenzie, H. A., and Murphy, W. H. (1970). *Biochim. Biophys. Acta*, **214**, 419.

61. Andrews, A. T., Taylor, M. D., and Owen, A. J. (1985). *J. Chromatogr.*, **348**, 177.

62. Davies, D. T. and Law, A. J. R. (1987). *J. Dairy Res.*, **54**, 369.

63. Guillou, H., Miranda, G., and Pelissier, J. P. (1987). *Lait*, **67**, 135.

64. Manji, B., Hill, A., Kakuda, Y., and Irvine, D. M. (1985). *J. Dairy Sci.*, **68**, 3176.

65. Giradet, J. M., Paquet, D., and Linden, G. (1989). *Milchwissenschaft*, **44** (11), 692.

66. Laezza, P., Nota, G., and Addeo, F. (1991). *Milchwissenschaft*, **46** (9), 559.

67. Nichols, J. A. and Morr, C. V. (1985). *J. Food Sci.*, **50**, 610.

68. McNeill, G. P. and Donnelly, W. J. (1987). *J. Dairy Res.*, **54**, 19.

69. Anderson, M., Griffin, M. C. A., and Moore, C. (1984). *J. Dairy Res.*, **51**, 615.

70. Robson, E., Horne, D. S., and Dalgleish, D. G. (1985). *J. Dairy Res.*, **52**, 391.

71. Anderson, M., Moore, C., and Griffin, M. C. A. (1986). *J. Dairy Res.*, **53**, 585.

72. Barrefors, P., Ekstrand, B., Fagerstam, L., Larsson-Raznikiewicz, M., Schaar, J., and Steffner, P. (1985). *Milchwissenschaft*, **40**, 257.

73. Mackey, K. L., Cooke, P. H., Malin, E. L., and Holsinger, V. H. (1991). *J. Dairy Sci.*, **74**, 128 (abstract D143).

74. Yoshida, S. (1988). *J. Dairy Sci.*, **71**, 1.

75. Al-Mashikhi, S. A. and Nakai, S. (1987). *J. Dairy Sci.*, **70**, 2486.

76. Addeo, F., Chobert, J.-M., and Ribadeau-Dumas. B. (1977). *J. Dairy Res.*, **44**, 63.

77. Lindahl, L. and Vogel, H. J. (1984). *Anal. Biochem.*, **140**, 394.

78. Sanogo, T., Paquet, D., Aubert, F., and Linden, G. (1989). *J. Dairy Sci.*, **72**, 2242.

79. Haggarty, N. (1993). (private communication).

80. Sciancalepore, V., Pizzuto, P., and De Stefano, G. (1994). *Latte*, **19** (11), 1152.

81. Musznska, G., Dobrowolska, G., Medin, A., Ekman, P., and Porath, J. O. (1992). *J. Chromatogr.*, **604** (1), 19.

82. Pearce, R. J. (1983). *Austr. J. Dairy Tech.*, **38**, 114.

83. Lukien, J., Van der Zee, R., and Welling, G. W. (1984). *J. Chromatogr.*, **284**, 482.

84. Whitney, R. McL. (1988). In *Fundamentals of dairy chemistry* (ed. N. P. Wong), p. 81. Van Nostrand Reinhold Company, New York.

85. Siebert, B., Erhardt, G., and Senft, B. (1985). *Anim. Blood Groups Biochem. Genet.*, **16**, 183.

86. Vegarud, G. E., Molland, T. S., Brovold, M. J., Devold, T. G., Alestrom, P., Steine, T., *et al.* (1989). *Milchwissenschaft*, **44** (11), 689.

87. Bovenhuis, H. and Verstege, A. J. M. (1989). *Netherlands Milk Dairy J.*, **43**, 447.

88. de Jong, N., Visser, S., and Olieman, C. (1993). *J. Chromatogr.*, **652**, 207.

89. Paterson, G. R., Hill, J., and Otter, D. E. (1995). *J. Chromatogr.*, **700**, 105.

90. Otte, J., Zakora, M., Kristiansen, K. R., and Qvist, K. B. (1997). *Lait*, **77**, 241.

91. Leonil, J., Mollé, D., Gaucheron, F., Arpino, P., Guénot, P., and Maubois, J. L. (1995). *Lait*, **75**, 193.

92. Chen, J. P. and Wang, C. H. (1991). *J. Food Sci.*, **56**, 701.

93. Eigel, W. N., Butler, J. E., Ernstrom, C. A., Farrell, Jr. H. M., Harwalker, V. R., Jenness, R., *et al.* (1984). *J. Dairy Sci.*, **67**, 1599.

94. Fox, P. F. (1982). In *Developments in dairy chemistry. Vol. I. Proteins.* (ed. P. F. Fox), p. 189. Applied Science Publishers, London.

95. Fasman, G. D. (1989). In *CRC practical handbook of biochemistry and molecular biology*. CRC Press, Boca Raton, Florida.

96. Chemical Rubber Company. (1970). In *CRC handbook of biochemistry*, 2nd edn. The Chemical Rubber Co., Cleveland, Ohio.

97. Walstra, P. and Jenness, R. (1984). In *Dairy chemistry and physics*. John Wiley, New York.

98. McKenzie, H. A. and Murphy, W. H. (1970). In *Milk proteins. Vol. I* (ed. H. A. McKenzie), p. 127. Academic Press, New York.

99. Fox, P. F. (1992). In *Advanced dairy chemistry. Vol. 1. Proteins* (ed. P. F. Fox). Elsevier Applied Science, London.

100. Sigma Chemical Company, St. Louis, Mo.

101. Swaisgood, H. A. (1982). In *Developments in dairy chemistry. Vol. 1. Proteins* (ed. P. F. Fox), p. 1. Elsevier Applied Science, London.

Chapter 7

Protein purification from animal tissue

Nigel M. Hooper

School of Biochemistry and Molecular Biology, University of Leeds, Leeds LS2 9JT, UK.

1 Introduction

Numerous proteins have been purified from almost every conceivable animal tissue. In this chapter, I deal with the principles involved in isolating proteins from animal tissue, covering such issues as the choice of tissue, methods used to disrupt the tissue, and procedures for isolating the appropriate subcellular fraction. Most of the procedures that I detail have been used to isolate mg quantities of a particular protein from relatively large (50–200 g) amounts of tissue, but can readily be scaled down for smaller amounts. Also covered are the use of protease inhibitors to prevent unwanted proteolysis, and the use of detergents, proteases, and phospholipases to solubilize membrane proteins. The chapter finishes with three example purifications of animal proteins—the isolation of angiotensin converting enzyme and aminopeptidase P from pig kidney cortex membranes, and of fructose-1,6-bisphosphatase from pig liver cytosol.

2 Choice of tissue

Several factors will determine the choice of starting tissue from which to purify the protein of interest. Obviously, the ideal tissue will be one in which the protein is abundant. Thus, for example, a number of the cell-surface peptidases are present at high levels in the kidney cortex (1), and as this organ is relatively large, significant amounts of protein can often be isolated, e.g. 10–15 mg of angiotensin converting enzyme can be purified from 200 g of kidney cortex (2). However, when isolating a protein for the first time when limited information is available on its tissue distribution the most abundant source may not be obvious. Such was the case with the phospholipase D activity that cleaves glycosyl-phosphatidyl-inositol (GPI) membrane anchors. This activity was originally described in membrane fractions from numerous mammalian tissues. However, subsequently the enzyme was shown to be an abundant soluble protein in plasma (3), with the reported tissue activity probably due to contamination by plasma during the isolation procedure.

The second factor to take into account is the availability of the animal tissue. Local abattoirs can be useful sources of most tissues from larger domestic animals (pigs and cows in particular), whereas tissues from smaller animals (mice, rats, rabbits, etc.) can usually be obtained from institutional animal facilities or a commercial supplier. Obviously mouse tissues may be suitable for purifying small quantities of a protein, whereas a porcine liver would be more suitable for larger quantities. Generally speaking human tissue is more difficult to obtain, although we have often used term placentae which are readily available from the local maternity hospital. Nowadays good laboratory practice should be used when handling any animal tissue, with a stricter code of practice being employed when handling unscreened human tissue, i.e. that which has not been screened for AIDS and hepatitis B (see Appendix at the end of this chapter). In addition, with the current controversy over the source and route of transmission of the transmissible spongiform encephalopathies such as 'mad cow disease' and newer strains of scrapie, it is probably advisable to use similar precautions when handling bovine and sheep tissue.

Other factors to take into account include the freshness of the tissue and the ease with which it can be disrupted (see Section 3). For the purification of some proteins it may be essential to use fresh rather than frozen tissue, and this again will affect the source as well as the purification strategy.

3 Tissue disruption

Once an appropriate tissue has been chosen as the starting material, the next step is to disrupt its structure. The method of disruption of the tissue may depend on whether the protein is intracellular or extracellular, soluble or membrane-bound, or located in a subcellular organelle. With a large organ such as a pig kidney or human placenta the first thing to do is to cut it into approx. 1 cm^2 pieces using a pair of scissors or a scalpel. For the subsequent homogenization of large amounts of tissue a Waring blender with a capacity of up to 1.5 litres provides a means of rapidly disrupting large amounts of tissue. For smaller amounts of tissue alternative homogenizers can be used (e.g. Ultratirrax, Polytron, Braun). For soft tissue Potter-Elvehjem or Dounce glass–glass or glass–Teflon homogenizers can be used. A suitable homogenization buffer for most tissues is 50 mM Hepes/NaOH pH 7.4 containing isotonic (0.33 M) sucrose. The homogenization and subsequent subcellular fractionation (see Section 5) should be carried out at 4°C with pre-chilled buffers to minimize unwanted proteolysis. In addition, depending on the protein, protease inhibitors may need to be included in the homogenization buffer (see Section 4). Some tissues homogenize much more easily than others, depending to a large extent on the amount of connective tissue that they contain. Thus lung tissue is considerably more difficult to homogenize than brain cortex or kidney. Following homogenization, large pieces of tissue remaining can readily be removed either by straining the homogenate through muslin or fine gauze or by centrifugation at low speed (see

Section 5). Homogenization times should be kept to a minimum in order to minimize thermal denaturation of the proteins.

4 Prevention of unwanted proteolysis

Some proteins are particularly susceptible to proteolytic degradation whereas others are very resistant. As a general rule those proteins that are present on the cell-surface or are secreted in soluble form and which are glycosylated tend to be more resistant to proteolytic degradation. As soon as the tissue is homogenized proteases that are normally in a different subcellular compartment (in particular those in the lysosomes) will be liberated into solution and come into contact with the protein of interest, possibly degrading it. Performing all the homogenization and subsequent fractionation procedures at 4°C with pre-chilled buffers goes some way to minimizing unwanted proteolysis. Further control of proteolysis can be obtained by including protease inhibitors in the various buffers. *Table 1* provides a list of commercially available protease inhibitors. We have used the following two inhibitor cocktails to prevent unwanted proteolysis during the isolation of mammalian proteins:

(a) EDTA (5 mM), Dip-F (0.1 mM), E-64 (10 μM), antipain (10 μg/ml), leupeptin (10 μg/ml), pepstatin A (1 μM).

(b) EDTA (1 mM), 1,10-phenanthroline (1 mM), Dip-F (0.1 mM), PMSF (1 mM), E-64 (10 μM), and pepstatin A (1 μM).

One of the key factors to take into consideration when using such inhibitors is the cost, especially if repeatedly isolating a protein on a large scale. Another factor to bear in mind is that when isolating a protease and monitoring its purification by assaying for enzyme activity, it is prudent to avoid using inhibitors of the class of protease to which it belongs! Other lists of protease inhibitors and their properties can be found in Chapter 1 of the companion volume in this series (4), and a useful leaflet titled 'The easy way to customize your protease inhibitor cocktail' produced by Calbiochem.

5 Subcellular fractionation

Once the tissue has been homogenized the next stage is to isolate the subcellular fraction in which the protein of interest is located. This is most commonly achieved by differential centrifugation. *Figure 1* shows a general scheme for the subcellular fractionation of animal tissue which was originally described for use on rat kidney (5) but which we have used for several other tissues (6). *Protocol 1* details this subcellular fractionation procedure. If purifying a protein that resides within a subcellular organelle, such as pyruvate dehydrogenase from mitochondria (7), an additional step is required to break open the organelle. This could involve sonication, osmotic lysis, or homogenization in a Dounce homogenizer depending on the subcellular organelle and the properties of the pro-

Table 1 Properties of class specific protease inhibitors

Class of protease inhibited	Inhibitor	Mode[a]	Stock concentration	Effective concentration	Notes[b]
Metallo	EDTA	R	50 mM	1–5 mM	
	1,10-phenanthroline	R	100 mM	1 mM	Dissolve in methanol.
Serine	Aprotinin	R	1–10 mg/ml	2–10 µg/ml	
	Di-isopropylfluorophosphate (DipF)	I	1 M	0.1 mM	*Extremely toxic.* Dissolve in propan-2-ol. Half-life in aqueous solution 1 h at pH 7.5.
	Phenylmethylsulfonyl fluoride (PMSF)	I	100 mM	1 mM	Dissolve in propan-2-ol and store at 4°C. Half-life in aqueous solution 1 h at pH 7.5.
	3,4-Dichloroisocoumarin (3,4-DCI)	I	10 mM	5–100 µM	Dissolve in DMSO. Half-life in aqueous solution 20 min at pH 7.5.
Cysteine	E-64 (trans-epoxysuccinyl-L-leucylamido-(4-guanidino)butane)	I	1 mM	10 µM	
	Iodoacetamide	I	100 mM	1 mM	Prepare fresh as required.
	p-Hydroxymercuriphenyl sulfonic acid	I	100 mM	1 mM	
Serine/cysteine[c]	Antipain	R	1 mM	1–10 µM	Dissolve in methanol.
	Chymostatin	R	10 mM	10–100 µM	Dissolve in DMSO.
	Leupeptin	R	10 mM	10–100 µM	
Acidic	Pepstatin A	R	1 mM	1 µM	Dissolve in methanol.

[a] The mode of inhibition is indicated as either reversible (R) or irreversible (I).

[b] Unless otherwise indicated (i) the inhibitor should be dissolved in water to the indicated stock concentration and then stored at −20°C, (ii) the stock solution is stable when stored at −20°C for up to six months, and (iii) the inhibitor is available from Sigma Chemical Company.

[c] Trypsin-like serine proteases and some cysteine proteases.

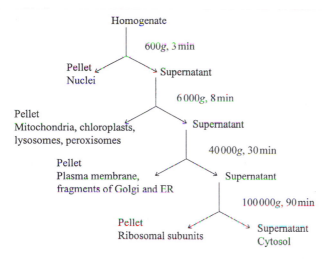

Figure 1 Scheme for the subcellular fractionation of porcine kidney.

tein. Further purification of a particular fraction obtained from the procedure detailed in *Protocol 1* can be achieved by density gradient centrifugation using sucrose or Percoll gradients. Details of methods to isolate nuclei and nuclear membranes, mitochondria, lysosomes, peroxisomes, and the Golgi apparatus can be found in a companion volume in this series (8). However, the relatively small amount of material that can be effectively separated in this manner usually precludes density gradient centrifugation from being used during the routine purification of proteins.

Protocol 1

Subcellular fractionation of porcine kidney

Equipment and reagents
- Homogenization buffer: 50 mM Hepes/NaOH pH 7.5, 0.33 M sucrose
- Ultracentrifuges with appropriate rotors

Method

1   Homogenize 20 g of porcine kidney cortex in 10 volumes (200 ml) of homogenization buffer.[a]

2   Centrifuge the homogenate at 600 g for 2.5 min.

3   Remove the supernatant from above the pellet of nuclei and cell debris[b] and centrifuge it at 6000 g for 8 min.

4   Remove the supernatant from above the lysosomal and mitochondrial pellet[b] and centrifuge it at 40 000 g for 30 min.

Protocol 1 continued

5 Remove the supernatant from above the microsomal pellet[b] and centrifuge it at 100 000 g for 90 min.

6 Finally remove the cytosolic fraction from above the light microsomal and ribosomal pellet.[b]

[a] The blender, and the homogenization and resuspension buffers should be pre-chilled to 4 °C.

[b] The pellet fractions should be resuspended in appropriate buffers as required.

For the large scale purification of plasma membrane-bound proteins we use a simplified differential centrifugation procedure (*Protocol 2*) that can accommodate a large volume (1.5 litres) of homogenate in the GSA/SLA1500 Sorvall rotors. Although a crude microsomal membrane fraction is obtained this is quite suitable for the further isolation of membrane-associated proteins. The supernatant obtained after centrifugation at 26 000 g (*Protocol 2*) can either be used directly or centrifuged further at 100 000 g for 1 h and then used as the source of soluble cytosolic proteins.

Protocol 2

Isolation of a crude microsomal membrane fraction from porcine kidney

Equipment and reagents

- Homogenization buffer: 50 mM Hepes/NaOH pH 7.4, 0.33 M sucrose
- Resuspension buffer: 10 mM Hepes/NaOH pH 7.4

- Preparative ultracentrifuge and rotors (e.g. Sorvall RC-5B refrigerated centrifuge, GSA/SLA1500 and SS34 rotors)

Method

1 In a Waring blender homogenize 100 g of porcine kidney cortex in 10 volumes (1 litre) of homogenization buffer.[a]

2 Centrifuge the homogenate at 8000 g for 15 min.

3 Carefully decant off the supernatant and centrifuge it at 26 000 g for 2 h.

4 Resuspend the resulting microsomal pellet in resuspension buffer to a final concentration of approx. 10 mg of protein per ml. This resuspended microsomal pellet is then used as the source of membrane-bound proteins.

[a] The blender, and the homogenization and resuspension buffers should be pre-chilled to 4 °C.

The apical microvillar or brush-border membrane of kidney and intestinal epithelial cells is enriched in numerous ectoenzymes including alkaline phosphatase, 5'-nucleotidase, γ-glutamyl transpeptidase, angiotensin converting enzyme, and numerous other peptidases (1). Aggregation of other subcellular structures with either Mg^{2+} or Ca^{2+} (*Protocol 3*) allows a membrane fraction highly enriched in the microvillar membrane, and essentially free of contaminating membranes, to be isolated (9–11).

Protocol 3

Isolation of a microvillar membrane fraction from porcine kidney

Equipment and reagents

- Preparative ultracentrifuge and rotors (e.g. Sorvall RC-5B refrigerated centrifuge, GSA/SLA1500 and SS34 rotors)
- Solid $MgCl_2.6H_2O$
- Homogenization buffer: 10 mM mannitol, 2 mM Tris–HCl pH 7.1
- 10 mM Hepes/NaOH pH 7.4

Method

1 In a Waring blender homogenize 50 g of porcine kidney cortex in homogenization buffer to give a 10% (w/v) homogenate.[a]

2 Place the homogenate in a beaker on ice and add 1.0 g $MgCl_2.6H_2O$ with stirring. Leave the suspension on ice for 15 min with occasional stirring.

3 Centrifuge the suspension at 1500 g for 12 min to pellet other subcellular structures.

4 Decant off the supernatant into a clean centrifuge tube and centrifuge it at 15 000 g for 15 min to pellet the microvillar membrane fraction.

5 Carefully remove the second supernatant from above the loose pale pink pellet with a Pasteur pipette. Resuspend the pellet in 50 ml of homogenization buffer using a vortex mixer.

6 To the resuspended pellet add a further 0.5 g of $MgCl_2.6H_2O$ with stirring. Leave the suspension on ice for 15 min with occasional stirring as before.

7 Centrifuge the suspension at 2200 g for 12 min to pellet other subcellular structures.

8 Decant off the supernatant and centrifuge this at 15 000 g for 15 min.

9 Finally, carefully remove the supernatant and resuspend the brush-border membrane pellet in approx. 25 ml of 10 mM Hepes/NaOH pH 7.4.

[a] The blender, and the homogenization and resuspension buffers should be pre-chilled to 4 °C.

Soluble extracellular proteins such as those found in plasma or other body fluids can readily be obtained by pelleting the cells by centrifugation and then removing the supernatant containing the soluble proteins. *Protocol 4* details a procedure for the fractionation of fresh porcine blood, but which is applicable to other animals, where a citrated glucose solution is used to prevent the blood from clotting.

Protocol 4

Fractionation of porcine blood

Equipment and reagents

- Citrated glucose solution: 2.51 g citric acid (72 mM), 4.18 g sodium citrate (85 mM), 3.34 g glucose (111 mM) dissolved in 167 ml water
- Preparative ultracentrifuge and rotors
- 10 mM sodium acetate pH 5.4

Method

1 To 1 litre of fresh blood[a] add 167 ml of citrated glucose solution and mix by inversion.

2 Keep on ice until fractionation can be carried out and perform all subsequent steps at 4°C.

3 Centrifuge the blood at 1000 g for 10 min to pellet the cells.

4 Remove the supernatant to a clean tube and centrifuge it at 30 000 g for 20 min to pellet the platelets.

5 Remove the supernatant and dialyse it extensively against 10 mM sodium acetate pH 5.4.

6 Centrifuge the dialysate at 10 000 g for 30 min to precipitate the euglobulins.

7 Dialyse the resulting supernatant against a suitable buffer for the subsequent purification of the protein of interest.

[a] The citrated glucose solution should be added as soon as possible to the blood (within minutes of bleeding the animal) before it coagulates.

6 Solubilization of membrane proteins

If the protein of interest is membrane-bound it is necessary to solubilize it from the membrane prior to its purification. The method of solubilization will depend on how the protein is associated with the lipid bilayer. For those proteins with multiple membrane spanning regions or with covalently attached lipid moieties (e.g. myristate, palmitate, or prenyl groups) the membrane structure will have to be disrupted using a detergent. Numerous non-denaturing detergents are now available and the choice will depend on its effectiveness in solubilizing the protein of interest and maintaining it in an active conformation, its availability, and its cost. More details on membrane protein solubilization can be found in a companion volume in this series (12). Although Triton X-100 is a commonly used detergent for the solubilization of membrane proteins, its major limitation is its high UV absorbance making the monitoring of total protein elution from a column by measuring absorbance at 280 nm difficult. If a detergent is used to solubilize the membrane-bound proteins it is essential to include a detergent in all the buffers throughout the subsequent stages of the purification. Failure to do so will almost certainly result in the loss of the protein due to aggregation of the hydrophobic moieties and their non-specific binding to the surfaces of columns and vessels.

For Type I and Type II integral membrane proteins with a single membrane spanning region detergents can of course be used to solubilize them from the membrane. Alternatively, such proteins where the bulk of the polypeptide chain is located on one or other side of the bilayer, are often susceptible to cleavage in the membrane proximal stalk region by specific proteases such as trypsin or papain (2, 13, 14). The advantage of such limited proteolysis to release the protein from the membrane prior to its purification is twofold. First, protease cleavage generally solubilizes maximally 10–20% of the total membrane protein therefore providing a simple but effective initial purification step. Second, is that the resulting cleaved form of the protein is hydrophilic, negating the need for detergents to be included in the buffers during the subsequent purification steps.

GPI anchored proteins (15) can be effectively solubilized by those detergents with high critical micellar concentrations such as n-octyl-β-D-glucopyranoside, CHAPS, and deoxycholate, but are generally resistant to solubilization by detergents with a low critical micellar concentration such as Triton X-100 and Nonidet P-40 (16, 17). In addition, many mammalian GPI-anchored proteins are susceptible to release by bacterial PI-PLC which selectively cleaves the GPI anchor (18, 19). As with limited proteolytic cleavage, only a subpopulation of the membrane proteins are released by bacterial PI-PLC providing a useful initial purification step, and the released form of the protein is hydrophilic. More detailed information on the solubilization of membrane proteins with bacterial PI-PLC and the commonly encountered problems of this procedure can be found in a companion volume in this series (20). *Protocol 5* details how we routinely use detergents, proteases, or bacterial PI-PLC to solubilize mammalian membrane-bound proteins prior to their purification.

Protocol 5

Solubilization of membrane proteins

Equipment and reagents

- Preparative ultracentrifuge and rotor (e.g. Sorvall RC-5B refrigerated centrifuge, SS34 rotor)
- Triton X-100 (20%, w/v solution), or n-octyl-β-D-glucopyranoside, or trypsin, or bacterial PI-PLC[a]

Method

1 Incubate the resuspended microsomal membrane pellet (see *Protocol 2*) with one of the following:

   (a) Triton X-100 at a ratio of 7:1 20% Triton X-100 : protein (v/w) for 1 h at 4°C.

   (b) n-octyl-β-D-glucopyranoside at a final concentration of 60 mM for 1 h at 4°C.

   (c) Trypsin at a ratio of 1:10 trypsin : protein (w/w) for 1 h at 37°C.

   (d) Bacterial PI-PLC (0.1 unit per mg of protein) for 2 h at 37°C.

Protocol 5 continued

2 After the appropriate incubation time centrifuge the sample at 31 000 *g* for 90 min.

3 Decant off the resulting supernatant containing the solubilized proteins.

[a] PI-PLC from a number of species, particularly *Bacillus thuringiensis* and *B. cereus*, are available as recombinant proteins from a number of commercial suppliers. For large scale solubilization of GPI-anchored proteins we use *B. cereus* phospholipase C Type III (Sigma, Cat. No. P6135) which contains a contaminating PI-PLC activity. As the amount of the PI-PLC activity in this preparation can vary it is best to test each batch before use on a large scale (for further information see ref. 20).

Numerous methods are available for the removal of peripheral membrane proteins from the lipid bilayer which all essentially involve disrupting the electrostatic interactions between the protein and the phospholipid headgroups and/or other membrane proteins. Such procedures usually involve washing the membrane fraction with a high pH and/or a high salt containing buffer, for example 0.1 M Na_2CO_3 pH 11.0.

7 Example purifications of animal proteins

The above procedures should enable a soluble fraction containing the protein of interest to be isolated from the appropriate animal tissue. Further purification of the protein will depend on the particular protein and whether affinity chromatography or other selective chromatographic steps can be used. The remainder of this chapter details the purification of three proteins which provide examples of the types of isolation procedures that can be used on animal proteins. The first of these details the use of inhibitor affinity chromatography to purify angiotensin converting enzyme following its solubilization from the membrane with either Triton X-100 or trypsin. The second shows how a combination of conventional chromatographic steps were used to purify the GPI-anchored aminopeptidase P, while the third details the purification of the cytosolic fructose-1,6-bisphosphatase.

7.1 Purification of angiotensin converting enzyme

Angiotensin converting enzyme (EC 3.4.15.1) is a Type I integral membrane protein that also exists in a soluble form in plasma, and that plays a key role in the control of blood pressure (21). Inhibitors of this zinc metalloenzyme such as captopril, enalapril, and ramipril, are currently used clinically as anti-hypertensive agents. A derivative of enalapril, lisinopril has proved to be an excellent ligand for the affinity purification of this enzyme (22). Details of the procedure for coupling lisinopril to Sepharose with a long (2.8 nm) spacer arm can be found in ref. 23. *Protocol 6* details the affinity purification of angiotensin converting enzyme from porcine kidney cortex following isolation of a microsomal membrane fraction (*Protocol 2*) and solubilization of the enzyme from the membrane with limited trypsin treatment (*Protocol 5*). The solubilized protein sample is applied to the lisinopril affinity column which is linked in series behind a pre-column of unmodified Sepharose in order to reduce non-specific binding to, and prolong the life of, the affinity column. When purifying the amphipathic form of angiotensin converting enzyme that has been solubilized from the membrane with Triton X-100 (see *Protocol 5*) 0.1% Triton X-100 should be included in all the column buffers. This procedure can purify up to 15 mg of angiotensin converting enzyme in essentially a single step from 200 g of pig kidney cortex.

Protocol 6

Affinity purification of angiotensin converting enzyme on lisinopril–2.8 nm–Sepharose (23)

Equipment and reagents

- Peristaltic pump
- Glass or plastic columns
- Narrow bore (1–2 mm) tubing to connect the pump to the column
- Fraction collector

- Running buffer: 10 mM Hepes, 0.3 M KCl, 0.1 mM $ZnCl_2$ pH 7.5[a]
- Lisinopril–2.8 nm–Sepharose (23)
- Unmodified Sepharose CL-4B

Method

1 Pre-equilibrate a 1×6 cm (approx. 5–10 ml) column of lisinopril–2.8 nm–Sepharose and a 1.5×10 cm (approx. 10–20 ml) pre-column of unmodified Sepharose CL-4B with 200 ml of running buffer at a flow rate of 10–20 ml/h.

2 Apply the trypsin-solubilized protein sample (see *Protocol 5*) at a flow rate of 10–20 ml/h and collect run-through for determination of unbound protein.

3 Once all the sample has been applied, remove the pre-column and wash the affinity column with 400 ml of running buffer.

4 Elute the bound protein from the column by including 10 μM lisinopril in the running buffer or by using 50 mM sodium borate buffer pH 9.5. Collect 2 ml fractions and monitor the absorbance at 280 nm.

5 Pool those fractions with high absorbance at 280 nm[b] and dialyse extensively against 5 mM Tris–HCl pH 8.0.

6 Concentrate protein using centrifugal concentrators, e.g. Vivascience Vivaspin or Amicon Centricon.[c] Alternatively apply the protein to a small (0.5 ml) column of DEAE–cellulose pre-equilibrated with 5 mM Tris–HCl pH 8.0 and elute into 0.5–1.0 ml with the same buffer containing 0.7 M NaCl.

[a] $ZnCl_2$ is included here to maximize the binding of angiotensin converting enzyme to the lisinopril ligand.

[b] As only angiotensin converting enzyme should have bound and then been eluted, measurement of absorbance at 280 nm is the simplest method of determining which fractions contain the protein. If the buffers contain Triton X-100, alternative detection methods will be required as this detergent absorbs strongly at 280 nm.

[c] When the amphipathic form of angiotensin converting enzyme is isolated in the presence of detergent, we have found that the protein can bind irreversibly to the membrane in the centrifugal concentrators and thus recommend concentration on DEAE columns.

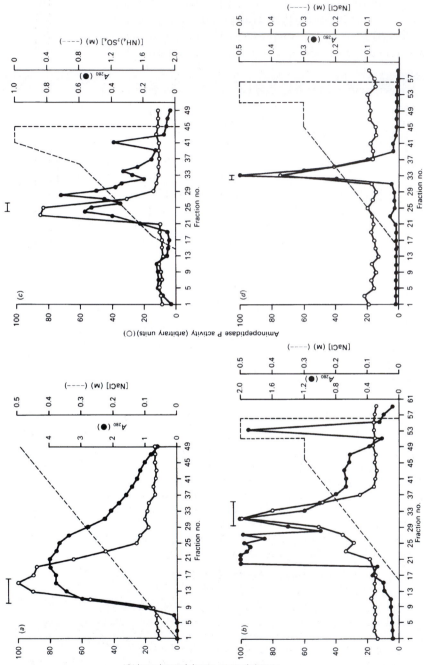

7.2 Purification of aminopeptidase P

Aminopeptidase P (EC 3.4.11.9) is a widely distributed mammalian cell-surface peptidase that has been implicated in the *in vivo* metabolism of bradykinin (24). This GPI-anchored protein was first purified following its release from the membrane by cleavage of its GPI anchor with bacterial PI-PLC (see *Protocol 5*) (19). The resulting hydrophilic form of the protein was then purified by a combination of anion exchange and hydrophobic interaction chromatographies as detailed in *Protocols 7–9*. The first anion exchange step on DEAE–cellulose not only provided a significant purification but also reduced the large sample volume resulting from the solubilization step to a small enough volume to allow FPLC to be used. After the hydrophobic interaction chromatography the sample was dialysed to remove the ammonium sulfate. Anion exchange chromatography on a MonoQ column was finally used to concentrate the aminopeptidase P and purify it to apparent homogeneity as assessed by sodium dodecyl sulfate polyacrylamide gel electrophoresis. *Figure 2* shows the results obtained for the purification of aminopeptidase P from porcine kidney cortex using these chromatographic steps. Fractions were pooled so as to maximize the purity of the enzyme at each stage rather than to maximize the recovery. Using this procedure 1.2 mg of aminopeptidase P was purified from 400 g of pig kidney cortex (19).

7.3 Purification of fructose-1,6-bisphosphatase

Fructose-1,6-bisphosphatase is a key control enzyme in gluconeogenesis which is present in the cytosol of liver cells. *Protocol 10* details the purification of the enzyme from porcine liver using ammonium sulfate precipitation and two cation exchange chromatography steps on CM–cellulose with different modes of elution; the second one employing substrate elution (25, 26). Dithiothreitol is present in all the buffers throughout the purification in order to maintain the

---

Figure 2 Purification of aminopeptidase P from porcine kidney cortex. Following solubilization of the microsomal membranes with bacterial PI-PLC (Protocols 2 and 5) the sample was applied to a DEAE–cellulose column (a) as detailed in Protocol 7. Bound protein was eluted with a linear gradient of 0–0.5 M NaCl, 5.0 ml fractions collected and assayed for enzyme activity and protein (A_{280}). Fractions 10–16 were pooled and dialysed against 10 mM Tris–HCl pH 7.6 before being applied to an HR5/5 MonoQ column (b). Bound protein was eluted with a non-linear gradient of 0–0.5 M NaCl, 1.0 ml fractions were collected and assayed for enzyme activity and protein (A_{280}). Fractions 29–35 were pooled and dialysed against 20 mM sodium phosphate pH 7.5 before being applied to an HR5/5 alkyl–Superose column (c) as detailed in Protocol 8. Bound protein was eluted with a non-linear gradient of 2.0–0 M $(NH_4)_2SO_4$, 1.0 ml fractions were collected and assayed for enzyme activity and protein (A_{280}). Fractions 24–26 were pooled and then chromatographed on a mixed affinity column of cilastatin– and lisinopril–Sepharose. The run-through from this column was finally dialysed against 10 mM Tris–HCl pH 8.0 and applied to an HR5/5 MonoQ column equilibrated in (d) as detailed in Protocol 9. Bound protein was eluted with a non-linear gradient of 0–0.5 M NaCl, 1.0 ml fractions were collected and assayed for enzyme activity and protein (A_{280}). Fractions 32 and 33 were pooled and used as purified aminopeptidase P. ○, aminopeptidase P activity; ●, protein. Reproduced with permission from Portland Press, London (19).

reducing environment of the cytosol, while PMSF is included in the initial homogenization buffer in order to minimize unwanted proteolysis. This procedure resulted in the purification of 2 mg of fructose-1,6-bisphosphatase from 50 g of pig liver.

Protocol 7

Large scale anion exchange chromatography on DEAE–cellulose

Equipment and reagents

- Peristaltic pump
- Glass or plastic column
- Narrow bore (1–2 mm) tubing to connect the pump to the column
- Fraction collector
- Running buffer: 10 mM Tris–HCl pH 7.6
- DEAE–cellulose (DE-52, Whatman)

Method

1 Perform all procedures at 4 °C. Extensively dialyse the PI-PLC solubilized sample (see Protocol 5) against the running buffer.

2 Centrifuge the sample at 31 000 g for 90 min to remove precipitated material.

3 Apply the resulting supernatant to a DEAE–cellulose column (30 ml bed volume) equilibrated in running buffer at a flow rate of 20–30 ml/h.

4 Once all the sample has been applied, wash the column with 1–2 column volumes of running buffer.

5 Elute bound protein with a 200 ml linear gradient of 0–0.5 M NaCl in running buffer and collect 5 ml fractions.

6 Measure the absorbance at 280 nm of each fraction and assay each for enzyme activity.

Protocol 8

Alkyl–Superose hydrophobic interaction chromatography

Equipment and reagents

- Pharmacia FPLC system
- Running buffer: 20 mM sodium phosphate pH 7.5
- Fraction collector
- HR5/5 alkyl–Superose FPLC column (Pharmacia)

Method

1 Extensively dialyse the sample against running buffer.

2 Mix the dialysed sample with an equal volume of 4 M $(NH_4)_2SO_4$ in running buffer and apply to an HR5/5 alkyl–Superose column equilibrated in 2 M $(NH_4)_2SO_4$ in running buffer.[a]

Protocol 8 continued

3   Elute bound protein with a gradient of 2.0–0.0 M $(NH_4)_2SO_4$ in running buffer and collect 0.5–1.0 ml fractions.

4   Measure the absorbance at 280 nm of each fraction and assay each for enzyme activity.

[a] Do not filter or centrifuge the sample once it has been mixed with the $(NH_4)_2SO_4$ as the precipitated protein will be lost.

Protocol 9

MonoQ anion exchange chromatography

Equipment and reagents

- Pharmacia FPLC system
- Fraction collector

- Running buffer: 10 mM Tris–HCl pH 8.0
- HR5/5 MonoQ FPLC column (Pharmacia)

Method

1   Extensively dialyse the sample against running buffer.

2   Apply the sample to an HR5/5 MonoQ FPLC column equilibrated in running buffer.

3   Elute bound protein with a gradient of 0.0–0.5 M NaCl in running buffer and collect 0.5–1.0 ml fractions.

4   Measure the absorbance at 280 nm of each fraction and assay each for enzyme activity.

Protocol 10

Purification of fructose-1,6-bisphosphatase

Equipment and reagents

- Preparative ultracentrifuge and rotors
- Peristaltic pump
- Glass or plastic columns
- Narrow bore (1–2 mm) tubing to connect the pump to the column
- Fraction collector
- CM–cellulose (Bio-Rad)

- Homogenization buffer: 0.15 M KCl, 0.1 mM DTT, 1 mM PMSF
- Running buffer A: 5 mM malonic acid/NaOH, 0.1 mM DTT pH 6.0
- Running buffer B: 5 mM malonic acid/NaOH, 0.1 mM DTT pH 6.5

Method

1   In a Waring blender homogenize 50 g of porcine liver in 200 ml of homogenization buffer. Then centrifuge the homogenate at 48 000 g for 1 h.

Protocol 10 continued

2 Remove the resulting supernatant and centrifuge it at 100 000 g for 1 h.

3 Remove the second supernatant and adjust it to pH 6.5 ± 0.1 by the addition of cold 5 M NaOH and then centrifuge again at 100 000 g for 1 h.

4 To each litre of this next supernatant slowly add 327 g $(NH_4)_2SO_4$ and stir for 30 min.

5 Centrifuge the suspension at 48 000 g for 20 min.

6 To each litre of the resulting supernatant slowly add 66 g of $(NH_4)_2SO_4$ and stir for a further 30 min.

7 Centrifuge the suspension at 48 000 g for 20 min.

8 Discard the supernatant and dissolve the pellet in 30 ml of running buffer A and dialyse extensively against the same buffer.

9 Load the dialysate onto a 2.5 × 10 cm column of CM–cellulose equilibrated in running buffer A at a flow rate of 30 ml/h.

10 After washing the column with 100 ml of running buffer A, elute bound protein with a 100 ml linear gradient of 5–30 mM malonic acid/NaOH, 0.1 mM DTT pH 6.0.

11 Pool those fractions with high fructose-1,6-bisphosphatase activity and concentrate using a Centriprep 10 concentrator into running buffer B.

12 Load the concentrated sample onto a 1 × 5 cm column of CM–cellulose equilibrated in running buffer B.

13 Wash the column with running buffer B until the eluate is protein-free as determined by absorbance measurement at 280 nm.

14 Elute bound protein with 0.1 mM fructose-1,6-bisphosphate in running buffer B and collect 1 ml fractions.

15 Pool those fractions with high enzyme activity and concentrate using a Centriprep 10 concentrator.

References

1. Hooper, N. M. (1993). In *Biological barriers to protein delivery* (ed. K. L. Audus and T. J. Raub), p. 23. Plenum Press, New York.

2. Hooper, N. M., Keen, J., Pappin, D. J. C., and Turner, A. J. (1987). *Biochem. J.*, **247**, 85.

3. Low, M. G. and Prasad, A. R. S. (1988). *Proc. Natl. Acad. Sci. USA*, **85**, 980.

4. Beynon, R. J. and Bond, J. S. (ed.) (1989). *Proteolytic enzymes: a practical approach*, p. 259. IRL Press, Oxford.

5. Shibko, S. D. and Tappel, A. L. (1965). *Biochem. J.*, **95**, 732.

6. Patel, D., Hooper, N. M., and Scott, C. S. (1993). *Br. J. Haematol.*, **84**, 608.

7. Lilley, K., Zhang, C., Villar-Palasi, C., Larner, J., and Huang, L. (1992). *Arch. Biochem. Biophys.*, **296**, 170.

8. Evans, W. H. (1987). In *Biological membranes: a practical approach* (ed. J. B. C. Findlay and W. H. Evans), p. 1. IRL Press, Oxford.

9. Booth, A. G. and Kenny, A. J. (1974). *Biochem. J.*, **142**, 575.

10. Kessler, M., Acuto, O., Storelli, C., Murer, H., Muller, M., and Semenza, G. (1978). *Biochim. Biophys. Acta*, **506**, 136.

11. Oppong, S. Y. and Hooper, N. M. (1993). *Biochem. J.*, **292**, 597.

12. Jones, O. T., Earnest, J. P., and McNamee, M. G. (1987). In *Biological membranes: a practical approach* (ed. J. B. C. Findlay and W. H. Evans), p. 139. IRL Press, Oxford.

13. Hesp, J. R. and Hooper, N. M. (1997). *Biochemistry*, **36**, 3000.

14. Hooper, N. M., Karran, E. H., and Turner, A. J. (1997). *Biochem. J.*, **321**, 265.

15. McConville, M. J. and Ferguson, M. A. J. (1993). *Biochem. J.*, **294**, 305.

16. Hooper, N. M. and Turner, A. J. (1988). *Biochem. J.*, **250**, 865.

17. Hooper, N. M. and Turner, A. J. (1988). *FEBS Lett.*, **229**, 340.

18. Littlewood, G. M., Hooper, N. M., and Turner, A. J. (1989). *Biochem. J.*, **257**, 361.

19. Hooper, N. M., Hryszko, J., and Turner, A. J. (1990). *Biochem. J.*, **267**, 509.

20. Hooper, N. M. (1992). In *Lipid modification of proteins: a practical approach* (ed. N. M. Hooper and A. J. Turner), p. 89. IRL Press, Oxford.

21. Williams, T. A., Soubrier, F., and Corvol, P. (1996). In *Zinc metalloproteases in health and disease* (ed. N. M. Hooper), p. 83. Taylor and Francis, London.

22. Bull, H. G., Thornberry, N. A., and Cordes, E. H. (1985). *J. Biol. Chem.*, **260**, 2963.

23. Hooper, N. M. and Turner, A. J. (1987). *Biochem. J.*, **241**, 625.

24. Turner, A. J. and Hooper, N. M. (1997). In *Cell-surface peptidases* (ed. A. J. Kenny and C. M. Boustead), p. 219. BIOS, Oxford.

25. Nimmo, H. G. and Tipton, K. F. (1982). In *Methods in enzymology* (ed. W. A. Wood), Vol. 90, p. 330. Academic Press, San Diego.

26. Heywood, S. P. (1996). ph.D. thesis, University of Leeds.

Appendix

Laboratory code of practice for the handling of unscreened human tissue

Note: This code of practice is currently used in the School of Biochemistry and Molecular Biology, University of Leeds. However, local and national regulations may necessitate further precautions to be taken in other work places.

Title of project: e.g. Purification of protein X from human tissue.

Staff member responsible:

Rooms used for experiments:

Names of experimenters:

Nature of samples: e.g. Post-mortem specimens, surgical biopsies, placental tissue.

Storage arrangements: e.g. Tissues and extracts will be frozen, labelled with Biohazard tape, and stored in designated freezers at -20 or $-70°C$.

Restrictions within rooms: e.g. Only bench so marked (i.e. unambiguous description of part of room in which the hazardous material is to be contained). Procedures involving dissection, homogenization, and transfer of supernatants or pellets after centrifugation we perform in a Class 2 microbiological safety cabinet.

Arrangements for decontamination and disposal: Include here arrangements for the disposal of gloves, aprons, masks, etc. and the sterilization of vessels and instruments prior to washing.

Experimental precautions: This will vary with the nature of the experiments. For example, procedures involving crude tissue samples and those that might generate aerosols (e.g. homogenization), or have a higher risk of spillage, we would contain within a Class 2 microbiological safety cabinet. Procedures involving clear, cell-free extracts subjected to normal protein purification would be performed on designated bench areas in the working laboratories. Protective gloves, masks, and aprons, as well as side or back fastening laboratory coats, would be worn for these procedures.

Chapter 8

Protein purification from plant sources

G. Paul Bolwell

School of Biological Sciences, Division of Biochemistry, Royal Holloway, University of London, Egham, Surrey TW20 0EX, UK.

1 Introduction

There is a prevailing perception that plants pose special problems with respect to protein purification. These problems occur as a result of lower protein per unit volume than animals or prokaryotes, the occurrence of natural products that interfere with purification and oxidative processes brought into play by tissue disruption that lead to enzyme inactivation. This chapter highlights how some of these problems may be overcome. Often a careful selection of source material and preliminary work to determine the time window of maximum expression of the target protein can improve the chances of purification to homogeneity significantly. Plants tend to regulate protein levels by changes in gene expression and turnover and these can often be in response to developmental and environmental factors. Prior knowledge of regulation of appearance of a particular protein is, particularly, advantageous. A second consideration, which is probably more important for plant sources, is the initial tissue disruption strategy and this requires some knowledge of the source material. Once the protein has been isolated in a form which allows fractionation however, plant proteins can be treated like any other and general chromatography matrices and methods are applicable.

The design of a purification protocol is of course dependent on the desired result. If protein sequence for design of oligonucleotides is the desired outcome, it may be that gel-based methods and blotting are sufficient if the target protein can be identified as a band. On the other hand, study of detailed enzyme kinetics will require a relatively pure enzyme since crude preparations may suffer from inhibitory protein–protein or protein–low molecular weight compound interactions or may represent mixtures of isoforms with differing kinetic properties. Similarly, if raising antibodies for immunological work on the protein is the goal then it will need to be purified to homogeneity. If the goal is crystallization, it should be noted that very few plant proteins have been crystallized after purification directly from the plant source which is ironic since the first protein ever crystallized, urease, was obtained from a plant source, jack bean. The usual route to crystallizing plant proteins is through cDNA cloning and heterologous expression followed by purification.

There have been some recent treatises on plant biochemistry which have focused on a large number of enzymes of primary and secondary metabolism (1, 2). In such surveys, it is striking that while many enzymes have been purified to homogeneity and the gene subsequently isolated, in many other cases it may be only just possible to assay an enzyme in a crude extract. Indeed there are extreme cases where a gene has been cloned, for example via a complementation of a mutant route (3) or through the identification of ESTs (expressed sequence tags) corresponding to bacterial homologues and using these to isolate the cDNA (4) where it has been difficult to prove the identity of the protein expressed in a heterologous system, enzymically. In these cases, complex dependencies of the system are not totally met and may duplicate the situation of some enzymes when assayed in crude form. In other cases, especially in secondary metabolism, the enzyme loses stereospecificity when isolated from the plant (5). This can sometimes be restored by the cooperation of other proteins (6). Finally, there may be restrictions due to the unavailability of substrates commercially, necessitating synthesis of these by the laboratory. Great strides in alkaloid metabolism have only been possible by this laborious route. This chapter is meant to give some very basic considerations when approaching plant systems. It is not meant to be comprehensive with respect to the types of assay systems and can only give a broad perspective about protein purification from plant sources. However, successful purification protocols for enzymes important and, in most cases unique to plant systems, are given inviting further research by the reader.

2 Special considerations

2.1 Low protein concentration

Unlike most other biological organisms, the bulk of the solid mass of plant material is not protein but other macromolecules such as polysaccharides, which include starch, cellulose, pectins and hemicellulose, and phenolic polymers. Furthermore, the initial homogenization has to be carried out in larger volumes of buffer to allow an initial filtration which would be hampered by any gelling effects or simple absorption back onto the large mass of insoluble material. As a consequence, there is a resultant initial low protein concentration which necessitates a precipitation step or use of a concentrator. Ammonium sulfate precipitation is the most commonly used step following homogenization as in most protocols, while PEG precipitation is also popular for plant proteins. Older methods used the preparation of acetone powders as the initial step to advantage but this has become less common over the years. If a concentrator is used then the ability of the plant protein to bind to the membrane should be considered as this can be a problem.

2.2 Proteolysis

Proteolysis is not always a particular problem with some plant sources. Where it presents itself, it is often compounded by the initial low concentration which may be exacerbated by the increased time spent on these initial steps before any chromatographic work which would reduce the degradation by separation of the protease from the target protein. If it is identified as a problem during empirical development of purification protocols it can be dealt with as in all protein purification by inclusion of a cocktail of protease inhibitors. It should be noted however that these may need modification for plants as they may have a different spectrum of protease types, necessitating a different spectrum of protease inhibitors. For instance, serine protease, so abundant in animal sources, may not be as abundant in the plant source. Nevertheless, inhibitors should be screened for effects on the target enzyme activity in crude extracts. Plants contain examples of the major classes of proteolytic enzymes, serine-, cysteine-, and metalloproteases, and the corresponding inhibitors can be used. One millimolar PMSF is used for serine proteases (*Table 1*). Thiol protease inhibitors such as *p*-chloromercuribenzoate and metalloprotease inhibitors such as 1,10-phenanthroline can cause problems with the target enzyme. Use of inhibitors such as pepstatin may be prohibitive in the initial homogenization on cost grounds due to large volumes but can be added after ammonium sulfate precipitation.

Table 1 Commonly used protectants in extraction buffers for plant tissues

Protectant[a]	Final concentration
Osmotica	
Sucrose	0.33–0.5 M
Sorbitol	0.33 M
Glycerol	10% (v/v)
Antioxidants	
2-Mercaptoethanol	1–10 mM
Dithiothreitol	1–10 mM
Ascorbate	1–20 mM
Phenolic oxidation inhibitors	
PVP	1–5% (w/v)
Amberlite XAD-2	1–5% (w/v)
Dowex 1×2 resin	1–5% (w/v)
Sodium metabisulfite	1–5 mM
Protease inhibitors	
Bovine serum albumin	1–5% (w/v)
PMSF	1 mM
p-Chloromercuribenzoate	1 mM
1,10-Phenanthroline	1 mM
EDTA	1 mM
Leupeptin	10–100 μM
Pepstatin	1 μM
Chymostatin	10–100 μM

[a] All these are commonly used reagents available from any general biochemical suppliers.

2.3 Inhibition and tanning

Many plant tissues are rich in oxidative enzymes that act upon endogenous substrates producing reactive products that react with and inactivate enzymes. These are probably part of the array of natural defence mechanisms of plants. Polyphenol oxidases are prominent in a number of species such as potato, apple, banana, and sugar beet and give rise to blackening reactions especially if free tyrosine is present. In other species such as legumes and monocots, peroxidases are the more abundant oxidative enzymes released by tissue disruption and give rise to browning reactions. There are number of strategies used alone or in combination to reduce or eliminate these undesirable reactions (*Table 1*). These are based on removal of the substrates or inhibition of the oxidative enzymes. Polyvinylpyrrolidone, used in the insoluble form if a soluble enzyme is being purified or the soluble form if membranes are desired, has been in long use for absorbing out the phenolic substrates for these enzymes. Cationic resins such as Dowex 1×2 or hydrophobic interaction resins such as Amberlite XAD-2 can also be used to remove substrates for these reactions. Inhibitors of the reactions such as thiols such as 2-mercaptoethanol or dithiothreitol are widely used. If these prove to have undesirable effects on the target enzymes then alternatives used are ascorbate (but watch for effects on homogenization buffer pH) or sodium metabisulfite.

3 Source of material

Whole plants are rarely used for protein purification. Organs such as leaf or stem are the usual primary source. However, rather rarely, it is possible to exploit particular plant structures in such a way to prepare relatively homogeneous cell types as starting material. This can have particular advantages.

3.1 Leaf

There is usually no problem with the abundance of leaf material. Also leaf cells have mainly primary walls and are relatively fragile. As a consequence, large volumes can be disrupted in a Waring blender, a Sorvall Omnimixer, or by using a Polytron homogenizer. It is useful to have a binocular microscope on hand (as in most cases) to monitor cell breakage which should be about 90%. Over homogenization will lead to plastid disruption which gives problems when purifying soluble enzymes. Leaves do not often pose much problems and are good sources. However, if an antibody is the desired outcome it is often made more difficult by contamination of the final product by ribulose bisphosphate carboxylase/ oxygenase (RUBISCO). Any final molecular weight for the purified protein of 57 kDa should be regarded with caution. In a similar way, and this applies to all plant sources, an M_r of between 60 and 65 kDa on SDS–PAGE for a low abundance protein should be regarded with suspicion as it might represent the well known gel artefact that arises if fresh sample buffer is not used to denature the sample. While blending methods are fine for most C_3 plants, C_4 plants such as maize, pose a special problem since mesophyll cells are more readily broken than bundle sheath cells. Complete extraction requires grinding methods to isolate all the photosynthetic enzymes from C_4 plants.

3.1.1 Glandular hairs and trichomes

These structures are often rich in particular proteins which afford purification of otherwise difficult enzymes. Two examples are polyphenol oxidase of potato leaf trichomes (7) and enzymes of monoterpene biosynthesis in glandular hairs of mint (8). The latter study has allowed spectacular success in allowing the purification and subsequent cloning of cDNAs encoding a number of difficult enzymes which probably would have not been possible without the initial cellular enrichment procedure (9). This involves gentle abrading of the leaf surface with glass beads (*Protocol 1*). Some 90% of the total leaf trichomes can be isolated in relatively pure and intact form using this method.

Protocol 1

Isolation of secretory cells from plant glandular trichomes

Equipment and reagents

- Cell disrupter (Bead Beater, Biospec Products) consisting of a 300 ml finned polycarbonate chamber fitted with a Teflon rotor which is sealed and attached to a motor
- XAD-4 resin (Amberlite, Rohm and Haas)
- Disruption buffer: 200 mM Hepes pH 7.3, containing 200 mM sorbitol, 10 mM sucrose, 5 mM dithiothreitol, 10 mM KCl, 5 mM $MgCl_2$, 0.5 mM potassium phosphate, 0.1 mM PPi (Fluka), 0.6% (w/v) methyl cellulose, and 1% (w/v) polyvinylpyrrolidone (PVP, M_r 40 000)

Method

1. Harvest apical buds and leaves and soak in ice-cold dH_2O for 1 h.
2. Abrade the tissue in the cell disrupter.
3. Add 15–20 g plant material together with 100–130 g glass beads (0.5 mm diameter) and XAD-4 resin and make up to full volume with disruption buffer.
4. Abrade leaves and buds using two to four pulses of operation of 1 min each with cooling on ice for 30 sec in between.
5. Control the rotor speed with rheostat settings between 75–90 V.
6. Collect secretory cells by successive filtration with washings in disruption buffer without PVP or methyl cellulose.
7. Pass the chamber contents through 350 μm nylon mesh to remove plant material, glass beads, and XAD-4 resin. Wash the trapped material and pass through the mesh several times.
8. Pass the filtrate through 105 μm nylon mesh to remove small tissue fragments and collect secretory cells by passing the 105 μm filtrate through 20 μm mesh.
9. This method can yield 90% of glandular hair tissue from the surface of the starting tissue.

3.2 **Stem**

Stem material suffers from low recovery of protein and their fibrous qualities which make tissue disruption difficult. For herbaceous plants, stems can be cut into small sections and ground in a mortar and pestle with an abrasive such as acid-washed coarse sand. It is sometimes useful to freeze first in liquid nitrogen and grind dry before addition of the buffer and formation of the slurry, providing the enzyme is stable at extreme temperatures. It is important to observe disruption in this case as it may be difficult to break many of the cells. The initial filtration step should involve rewashing of the insoluble material and re-filtration. One strategy to purify enzymes involved in wood formation from commercially important forest species has been to harvest logs during the onset of growth in the spring. These are then stored in a cold room for a short period. The bark can then be easily lifted and splits away at the cambial layer. This enables scraping away cell layers at different stages of vascular tissue development in quite high yields. As these are actively growing the protein content is high.

3.3 **Roots**

Roots are usually more fragile than stems but are very often quite rich in soluble phenolics. This is often dependent on growth conditions. Roots are sometimes also abundant in oxidative enzymes such as peroxidases (e.g. in maize), and tap roots and tubers (which are underground stems) in polyphenol oxidases (e.g. in sugar beet and potato). Routine precautions (PVP, Dowex, antioxidant, thiols, etc.) against oxidative tanning are usually sufficient to optimize purification from these sources.

3.4 **Seeds and storage organs**

Seeds and storage organs show a similar problem to leaves in that they contain high levels of storage proteins which can co-purify with the target protein and thus become very difficult to separate. Potato tubers are typical examples where the storage protein patatin is extremely abundant. Similarly, seed proteins are particularly abundant in legumes and monocots such as wheat and barley. Before embarking on protein purification from such sources it is useful to have prior knowledge of the M_r range of the abundant seed proteins. Any resultant purification of a target protein resulting in a similar M_r should be regarded with suspicion! Purification can also be aided if a step was included that effectively adsorbs out the major protein. Octyl-Sepharose was found to be effective in removing the major storage protein, patatin, during purification of polyphenol oxidase from potato tuber (10).

3.5 **Fruits**

Fruits are organs which undergo distinct developmental changes. The appearance of the target protein should be known and in the later stages of ripening a large number of degradative changes may be occurring in the tissues. Over-ripe fruit may be over-abundant in proteases and glycosidases giving rise to increased soluble polysaccharides which may contribute to precipitation of proteins in the initial clean-up filtration and centrifugation stages. Similar to leaf, fruits are often rich in plastids which subsequently undergo transition to chromoplasts. Green fruit can be left in the dark for several days before extraction to reduce contamination of plastid derived proteins released by breakage (P. F. Fraser, personal communication). Pigmented fruit do not pose any special problems due to the types of compounds present and are not particularly prone to oxidative processes during extraction.

3.6 Tissue cultures

Tissue cultures can be an abundant source of starting tissue. They provide a relatively uniform cell type. Although autotrophic cell cultures are feasible most bulk cultures are grown heterotrophically and are thus depleted in chloroplasts avoiding some of the problems encountered with green tissues. They are also often depleted in enzymes and compounds which cause problems in whole plant sources. They have proven useful sources for the purification of enzymes of secondary metabolism. Although most cell cultures approximate to meristematic cells produced upon wounding and thus distinguished by their primary walls, there are a number of manipulations possible that allow switching on of pathways. This has allowed the purification of enzymes not normally accessible from whole plant tissue because of abundance and inactivation. An extremely useful method of manipulation is elicitation. In this process, material of fungal origin is added to the culture, switching on distinct secondary pathways characteristic of the species. This has allowed purification of enzymes of phenolic, monoterpene, and alkaloid metabolism in particular. Some elicitors, such as that from bakers yeast are reasonably universal in their application, other elicitors are species-specific and model particular specific plant pathogen interactions. Tissue cultured cells can be grown in bulk, easily harvested, and large amounts of cells can be conveniently disrupted using a Polytron homogenizer.

3.7 Organelles

Plant cells have several structures which are predominantly associated with them, the chloroplast, vacuole, and the cell wall. Certain other structures such as the endomembrane system show structural differences in the level of proliferation of ER and Golgi which distinguishes them from other cell types. These can be sources in themselves and expedite the purification process. The cell wall is a product of the secretory system and because of the unique nature of the cell wall components, ER, Golgi, and vesicles contain enzymes characteristic of plants alone.

3.7.1 Chloroplasts and chromoplasts

There are many published methods for the preparation of intact and active chloroplasts. However, most of the published methods for the isolation of photosynthetic enzymes do not include a chloroplast isolation step but proceed directly with ammonium sulfate precipitation from filtered homogenates. There are other enzymes involved in other pathways that are chloroplast localized for which a chloroplast enriched preparation is an essential step. If this is required the chloroplasts may be isolated in a variety of osmotica of which 0.33 M sorbitol is most commonly used. Sucrose or mannitol are possible substitutes. Similarly, the buffer is not critical and Tris, MES, and Hepes are widely used. Common additives are $MgCl_2$ and $MnCl_2$ as well as inorganic phosphate and EDTA. Isoascorbate and BSA are the usual protective agents. Leaves are blended for a short period, filtered, and the filtrate subject to a low speed spin (e.g. 1000 g). The intact chloroplasts are resuspended by gentle stirring or using a soft paintbrush. The chloroplasts can be further purified on density gradients which can be sugar (e.g. sucrose), polysaccharide (e.g. Ficoll), or inorganic (e.g. Percoll)-based. A typical discontinuous sucrose gradient is given in *Protocol 2*. Fruit chromoplasts have also been isolated on density gradients. However for chromoplasts from Capsicum, sucrose was found to be the only suitable osmoticum and the organelles band at the 0.84/1.45 M interface (*Protocol 2*). Crude or purified plastid preparations can then be disrupted in a suitable buffer of which 100 mM Tris–HCl pH 7.8 containing 1 mM EDTA, 10 mM $MgCl_2$, 15 mM 2-mercaptoethanol, and 0.05% Triton X-100 has wide general use.

Protocol 2

Isolation of chloroplasts and chromoplasts for enzyme purification

Equipment and reagents

- Waring blender
- Muslin
- Discontinuous sucrose gradient for chloroplasts: 0.75, 1, 1.5 M sucrose (10 ml per fraction) in 50 mM Tris–HCl pH 7.6 containing 1 mM 2-mercapto-ethanol

- Isolation medium: 50 mM Tris–HCl pH 8 containing 1 mM EDTA, 0.4 M sucrose, 1 mM 2-mercaptoethanol
- Discontinuous sucrose gradient for chromoplasts: 0.45, 0.84, 1.45 M sucrose (9 ml per fraction) in 50 mM Tris–HCl pH 7.6 containing 1 mM 2-mercaptoethanol

Method

1   Chop deribbed leaves or green fruit (2 kg) or pigmented fruit (2 kg) into 2 litres of chilled isolation medium.

2   Disrupt the tissue twice for 1 sec each in a Waring blender and filter the slurry through four layers of muslin.

3   In all cases centrifuge the homogenate at 150 g for 5 min.

4   (a) For chloroplasts, centrifuge the supernatant from step 3 at 2200 g for 30 sec.

    (b) For chromoplasts, centrifuge the supernatant from step 3 at 2200 g for 10 min.

5 Discard the supernatants and for both cases, wash the crude organelles with isolation media, and pellet at the requisite centrifugation conditions.

6 Plastids are purified by sucrose density centrifugation.

7 (a) Resuspend the chloroplasts in 10 ml isolation medium, layer on a discontinuous sucrose gradient, and centrifuge at 1000 g for 15 min. Intact chloroplasts band at the 1/1.5 M interface; broken membranes at the 0.75/1 M interface.

(b) Resuspend the chromoplasts in 2 ml isolation medium, layer on a discontinuous sucrose gradient, and centrifuge at 62 000 g for 1 h. Intact chromoplasts band at the 0.84/1.45 M interface; broken membranes at the 0.45/0.84 M interface.

3.7.2 Microsomes

Purification of plant microsomal proteins can be difficult due to low abundance and restrictions on the volumes that can be handled. However, with the knowledge gained from the behaviour of membrane subfractions on sophisticated gradient centrifugation flotation methods can be used for enrichment. The first problem posed is that the high centrifuge speeds required to sediment microsomes usually mean the highest volume that can be handled is about 200 ml. However, Mg^{2+} precipitation is a useful 'trick' for obtaining bulk microsomes from large volumes. In this method, larger volumes of homogenate that have been centrifuged at 1000 g for 20 min to remove cell walls, chloroplasts, and nuclei are made to 25 mM by adding a suitable volume of 2 M $MgCl_2$. The supernatant is allowed to stand for 30 min at 4 °C to allow the microsomes to aggregate and these are then harvested by centrifugation at the maximum permissible speed for the size of rotor used (*Protocol 3*). This will allow pelleting of most of the ER and Golgi. Some membranes may not precipitate but often this is offset by the large increase in the amount of starting material which is a sufficient advantage for protein purification. Bulk microsomes prepared in this way can be further fractionated into preparations enriched in ER or Golgi membranes and this can be a useful enrichment step over a relatively crude microsomal starting position. Such a step can be most useful for Golgi-localized enzymes and dictyosome membranes band at the 0.5 M/1.0 M sucrose interface. It is not practical to use this approach for plasma membrane and two-phase methods are recommended in this case.

Microsomes can be readily solubilized. The author has found the use of hydrogenated Triton X-100 rather than non-reduced Triton X-100 to be a general detergent with wide applications for good recovery of activity following microsome solubilization. A suitable solubilization buffer would be 50 mM Tris–HCl containing 10 mM $MgCl_2$, 1 mM dithiothreitol, and 1% reduced Triton X-100. It is also an advantage to have the highest possible protein concentration. We have found using a bench-top ultracentrifuge such as a Beckman TL-100 allows solubilization in a small volume and simultaneous removal of particulate matter at high speed produces a supernatant which can be injected directly onto an HPLC system.

Protocol 3

Bulk preparation and fractionation of plant microsomes for protein purification

Equipment and reagents

- Homogenizer
- Muslin
- Homogenization medium: 50 mM Tris–HCl pH 7.6, 1 mM $MgCl_2$, 1 mM dithiothreitol, 0.33 M sucrose, 1% (w/v) PVP
- 2 M $MgCl_2$

- Resuspension medium: 50 mM Tris–HCl pH 7.6, 10 mM $MgCl_2$, 1 M dithiothreitol, 0.33 M sucrose
- Discontinuous gradient for microsomes: 0.5, 1.0, 1.6 M sucrose (10 ml per fraction) in 50 mM Tris–HCl pH 7.6 containing 1 mM 2-mercaptoethanol, 10 mM $MgCl_2$

Method

1 Homogenize plant material or tissue cultured cells in homogenization medium (1:1, w/v).

2 Filter the extract through three layers of muslin and centrifuge at 1000 g for 20 min.

3 Centrifuge the supernatant at 10 000 g for 10 min.

4 Add 2 M $MgCl_2$ to a final concentration of 25 mM and allow the microsomes to flocculate for 30 min at 4°C.

5 Harvest microsomes by centrifugation at the maximum speed for the size of rotor for 1 h.

6 Discard the supernatant and resuspend the pellet in resuspension medium.

7 Enrich the Golgi membranes by centrifugation of the crude microsomal fraction on a discontinuous sucrose gradient at 100 000 g for 1 h. The Golgi band is at the 0.5 M/1.0 M interface. The 1.0 M/1.6 M interface is enriched in ER.

3.7.3 Cell walls

i. Primary wall proteins

The plant cell wall is a dynamic system generally considered to be comprised of more than 90% carbohydrate polymers. Proteins, phenolics, and possibly lipids make up the remainder of the wall. This can have a bearing on the extractability of wall proteins. In fact much of the understanding of the range of wall proteins has come from gene cloning and has led to the identification of the glycine-, cysteine-, proline-, and hydroxyproline-rich subsets of structural wall proteins. In addition, many extracellular enzymes have been identified and are required for the restructuring and modification of this dynamic extracellular matrix in a variety of responses to developmental and environmental cues. However, there is a lack of direct studies on the proteins themselves and the true range of extracellular proteins and their species differences remains to be elucidated. It has recently been possible to perform a systematic extraction and sequencing of the major primary wall proteins from five species as members of four families of plants (11). This could be carried out by simple washing of tissue cultured cells (*Protocol 4*) and the subsequent N-terminal protein sequencing revealed many novel proteins. These washing steps can also be combined with subsequent chromatographic procedures to purify wall proteins to homogeneity (12, 13).

Protocol 4

Serial washes for plant primary cell wall proteins from tissue cultured cells

Equipment and reagents

- Miracloth
- 0.2 M CaCl$_2$
- 50 mM cyclohexane diaminotetraacetic acid (CDTA) (Sigma) in 50 mM sodium acetate pH 6.5
- 2 mM dithiothreitol (DTT)
- 1 M NaCl
- 0.2 M sodium borate pH 7.5

Method

1 Harvest cells for analysis four to five days after subculture by filtration on Miracloth. The culture filtrate was collected for further processing. Isolate wall protein subsets by the following sequential procedure.

2 Gently wash the cells washed with dH$_2$O and then stir in three volumes of 0.2 M CaCl$_2$ for 30 min.

3 Collect the cells and wash with several volumes of dH$_2$O. Then stir with three volumes of CDTA in 50 mM sodium acetate pH 6.5 after another wash with dH$_2$O.

4 Carry out the same procedure successively for 2 mM DDT, then 1 M NaCl, and finally 0.2 M sodium borate pH 7.5.

5 Filter all the extracts including the culture filtrate through GF/A paper before dialysis against dH$_2$O. Lyophilize all fractions before protein analysis.

ii. Secondary cell wall proteins

Secondary walls pose particular difficulties demanding harsher treatment than with primary walls (*Protocol 5*). However a successful protocol was designed to extract secondary wall specific proteins that carry a wheat germ agglutinin binding motif (14). Lectin chromatography (*Protocol 6*) is particularly useful for enriching certain subsets of extracellular glycoproteins. It has also been used for isolation from xylem cells scraped from de-barked sections of tree (15).

Protocol 5

Stronger treatments for extraction of secondary cell wall proteins

Equipment and reagents
- HB buffer: 50 mM Hepes pH 6.8 containing 2 mM EDTA, 5 mM 2-mercaptoethanol, and 0.4 M sucrose
- Homogenizer
- Nylon mesh

Method

1 Homogenize freshly collected stem tissue in HB buffer (2 ml/g fresh weight).

2 Pass the mixture through nylon mesh and wash the cell wall enriched pellet for 10 min with HB buffer with end-over-end mixing.

3 After centrifugation at 15 000 g for 15 min, resuspend the pellet in HB buffer containing 0.1% (v/v) Nonidet P-40 and incubate for 45 min with constant stirring.

4 Recover cell walls by centrifugation at 15 000 g for 12 min and wash three times, 10 min each, with deionized water containing 2 mM EDTA and 5 mM 2-mercapto-ethanol, centrifuging each time for 8 min at 10 000 g.

5 Extract the purified cell wall preparation with 1 M NaCl in HB buffer for 60 min and centrifuge at 15 000 g for 10 min.

6 Re-extract the pellet for 90 min with 1 M NaCl in HB buffer and repellet at 15 000 g for 10 min. Combine the high salt extracts and subject to further purification using lectin chromatography.

7 Make additional extraction steps with 1% (w/v) SDS in HB buffer at 70 °C for 1 h. Precipitate SDS-extracted proteins with ice-cold acetone, centrifuge at 25 000 g for 30 min, and store at −20 °C until further analysis.

8 Alternatively, extract with 1 M CaCl$_2$ in HB buffer for 1 h at 4 °C. Dialyse against several changes of deionized water for two days, freeze-dry, and store at −20 °C until further analysis.

Protocol 6

Lectin chromatography for extracellular glycoproteins

Equipment and reagents
- Concanavalin A (ConA)–Sepharose (Pharmacia or Sigma)
- Wheat germ agglutinin (WGA)–Sepharose 6MB (Sigma)
- TBS: 25 mM Tris–HCl pH 7.5, 0.15 M NaCl

Protocol 6 continued

Method

1 Precipitate salt-extracted proteins from primary or secondary walls by the addition of $(NH_4)_2SO_4$ up to 90% saturation.

2 After centrifugation, resuspend the resulting pellet in TBS, precipitate proteins with ice-cold acetone, and finally resuspended in and dialyse against TBS for 20 h.

3 Apply protein solution, cleared by centrifugation, to 7 ml column of WGA–Sepharose 6MB previously equilibrated with TBS, or ConA–Sepharose equilibrated with TBS containing 1 mM $MgCl_2$, 1 mM $CaCl_2$, and 1 mM $MnCl_2$.

4 After cycling phase, usually about 15 h, wash the column with 80 ml of TBS and elute proteins sequentially with:

   (a) For WGA–Sepharose
   - 25 ml of 0.1 M GlcNAc in TBS
   - 10 ml of 0.3 M GlcNAc in TBS
   - 10 ml of 0.3 M GlcNAc in 25 mM Tris–HCl pH 7.5, 0.5 M NaCl
   - 10 ml of 25 mM Tris–HCl pH 7.5, 0.5 M NaCl

   (b) For ConA–Sepharose
   - 25 ml of 0.1 M α-methyl mannoside in TBS containing 1 mM $MgCl_2$, 1 mM $CaCl_2$, and 1 mM $MnCl_2$
   - 10 ml of 0.3 M α-methyl mannoside in TBS, 1 mM $MgCl_2$, 1 mM $CaCl_2$, and 1 mM $MnCl_2$
   - 10 ml of 0.3 M α-methyl mannoside in 25 mM Tris–HCl pH 7.5, 0.5 M NaCl, 1 mM $MgCl_2$, 1 mM $CaCl_2$, and 1 mM $MnCl_2$
   - 10 ml of 25 mM Tris–HCl pH 7.5, 0.5 M NaCl

5 Combine fractions with a similar pattern of eluted proteins, dialyse against 25 mM NH_4HCO_3, and freeze-dry.

4 Direct and indirect strategies for identification of subsets of plant proteins

These strategies are defined as direct when fractionation is carried out by chromatography and the protein identified by a direct assay. The major specific considerations in plants are the tissue enrichment and disruption techniques. There are no specific considerations about chromatography and the same empirical nature about selection of matrices, running system (low pressure, FPLC, or HPLC), and elution buffers applies as for any other source. However, as already mentioned, plants are at the lower end of the scale as far as initial protein concentrations are concerned and careful considerations have to be made with respect to concentration steps and final recovery. As far as initial stages are concerned, ammonium sulfate precipitation is probably most commonly used while PEG precipitation is also popular. A desalting or clean-up stage then precedes the column chromatography protocol. Many purifications fail at the final recovery stage. This often incurs considerable losses due to proteins sticking to membranes used for concentrating and insufficient amounts to give a stable precipitate. The author's laboratory favours chloroform/methanol precipitation rather than a final ammonium sulfate or trichloroacetic acid precipitation (*Protocol 7*). However, if protein sequencing is the desired outcome then reverse-phase HPLC is recommended. The final product is assessed by SDS–PAGE and deemed to be pure if a single band revealed by silver staining is present. Sections 5 and 6 give protocols for the purification of typical plant proteins. This is not meant to be exhaustive but intended to give a flavour to the range of approaches and matrices adopted for proteins from plant sources.

Protocol 7

Methods for the final isolation step before protein sequencing

Two methods are used routinely in the author's laboratory.

A Chloroform/methanol precipitation

1   To one volume sample add four volumes methanol and mix.

2   Add one volume chloroform, mix.

3   Add one volume H_2O, mix, and centrifuge to form two phases.

4   Discard upper layer without disturbing the white protein layer at the interface.

5   Add three volumes methanol, mix, and centrifuge to pellet protein.

6   Remove supernatant and redissolve the protein in a suitable solvent.

Note: if the protein is to be sequenced then ultra pure chloroform must be used, impurities can cause N-terminal blocking.

B Reverse-phase HPLC

1   Dialyse the final protein sample against 0.1% (v/v) trifluoroacetic acid (TFA) and reduce in volume by partial lyophilization or in a concentrator.

2   Perform HPLC on an Aquapore RP 300 C8 column (Applied Biosystems) using a 0–70% gradient of 0.1% (v/v) TFA in 90% (v/v) acetonitrile.

3   Collect proteins by fraction collector or by hand. Proteins eluted in this way are entirely suitable for N-terminal protein sequencing.

However in many cases, loss of activity does not allow many column steps and even if this does not occur the final product may show multiple bands. At this stage indirect techniques for identification can be used. Gel electrophoresis methods can be used to tag the target enzyme. This can be carried out in native gels and involves some form of visual assay for product formation or labelling

Table 2 Purification of haem and copper oxidases showing assay system and chromatographic stages

Enzyme	Assay	1	2	3	4	5	6
CYP71 (19)	HP	SM	DE	GF			
CYP73 cinnamate 4-hydroxylase (20)	BS	SM	2P	DE	HA		
CYP74 allene oxide synthase (21)	HP	AP	AS	HI	MQ	MP	
CYP79 N-tyr hydroxylase (22)	BS	SM	DE/GFBS	RA			
CYP80 berbamunine synthase (23)	R	SM	HI	HA	MQ	MP	
Lignin peroxidase (12)	S	WW	LC	SE			
Laccase (13)	OE	WW	AS	LC	NG		
Polyphenol oxidase (10)	S	AS	DE	GF	LC	HI	GF

Key: BS, binding spectra assay; HP, hydroperoxide assay; OE, oxygen electrode assay; S, spectroscopic assay; R, radiometric assay; AS, ammonium sulfate precipitation; AP, acetone powder; SM, solubilized microsomes; WW, cell wall salt wash; AD, ADP–Sepharose; BS, Blue Sepharose; CF, chromatofocusing; DE, DEAE ion exchange; GF, gel filtration; HA, hydroxyapatite; HI, hydrophobic interaction chromatography; IF, isoelectric focusing; LC, ConA–Sepharose; MP, MonoP chromatofocusing (Pharmacia); MQ, Mono Q (Pharmacia); NG, non-denaturing gel electrophoresis; RA, Red Agarose; SE, size exclusion chromatography; 2P, two phase Triton X-114 separation.

the protein with a reactive substrate analogue, usually the azido derivative. The former approach has probably given best results and can be applied to oxidases (13), NAD(P)-requiring enzymes (16), and glycosyl transferases using the technique of product entrapment (17). Tagging with radiolabelled azido-derivatives often gives multiple banding patterns that are difficult to resolve and is probably best used as a last resort. One example where it has been used is to identify glucosyl transferases (18). [32P] UDP-glucose was used to identify a subunit of the cotton fibre glucan synthase.

A number of enzyme families are particularly prominent in plants and this can be used to advantage when a member of a particular subset is a target for purification. Haem-containing proteins can be purified by dual wavelength detection at 280 and 405 nm. This can be used for abundant haem proteins such as peroxidases and also difficult families such as the cytochrome P450s for which there is no direct assay since they require reconstitution with NADPH cytochrome P450 reductase. It is sometimes possible for these to be measured using hydroperoxides as donors or to be detected through binding spectra with substrates and inhibitors (19–23). This has led to purification of some cytochrome P450s (*Table 2*).

5 Examples of successful purification protocols: enzymes of carbohydrate metabolism

Plants are the most notable exponents of carbohydrate metabolism. In addition to the catabolic pathways found in all eukaryotes, they also fix carbon and are abundant producers of polysaccharides, both structural and storage.

5.1 Calvin cycle (24)

Paradoxically, despite the fact that all the genes have been cloned for Calvin cycle enzymes, not all the enzymes have been purified to homogeneity (*Table 3*). Some of these enzymes have cytosolic and plastid forms which need to be distinguished and separated. Most of the methods do not start with chloroplast preparations and proceed directly from whole cell homogenates.

5.2 Polysaccharide synthases

5.2.1 Starch and sucrose (25)

The enzymes of starch synthesis, ADP glucose pyrophosphorylase, starch synthase, and branching enzyme form a relatively short pathway but are important commercial targets. For purification purposes they are treated as soluble en-

Table 3 Successful protocols for the purification of the enzymes of the Calvin cycle and C_4 photosynthesis (24)

Enzyme	Assay	1	2	3	4	5
Rubisco	R	DG	MQ			
Fructose bisphosphatase	S	AS	IE	GF		
Sedoheptulose bisphosphatase	S	CS	DE	GF	BS	
Ribulose-5-phosphate kinase	S	AS	HI	BS		
PEP carboxylase	S	PG	IE	GF	BS	SE
Pyruvate dikinase	S	PG	IE	GF	BS	
PEP carboxykinase	S	AS	DE	DE		

Key: R, radiometric assay; S, spectrophotometric assay; AS, ammonium sulfate precipitation; CS, chloroplast stroma; DG, density gradient centrifugation; PG, PEG precipitation; BS, Blue Sepharose; DE, DEAE ion exchange; GF, gel filtration; HI, hydrophobic interaction chromatography; MQ, Mono Q (Pharmacia); SE, size exclusion chromatography. Where size exclusion is specified rather than gel filtration stipulates an HPLC separation system.

Table 4 Protocols for the purification of enzymes of starch and cell wall polysaccharide biosynthesis

Enzyme	Assay	1	2	3	4	5
UDPglc pyrophosphorylase (25)	R	PG	DE	HI	HI	
Starch synthase (25)	R	SG	DE	HI		
Starch branching enzyme (25)	R	SG	DE	HA	GF	
Sucrose phosphate synthase (26)	R	PG	DE	HI	MQ	HA
UDPglc dehydrogenase (16)	A	AS	HI	GF	BS	
Arabinan synthase (27)	R	SM	DE	SE	RP	
Callose synthase (28)	R	SM	DE	SE		
Xylan synthase (27)	R	SM	DE	SE		
Fucosyl transferase (17)	R	SM	GF	NG	IF	

Key: R, radiometric assay; S, spectrophotometric assay; AS, ammonium sulfate precipitation; PG, PEG precipitation; SG, starch grain isolation; SM, solubilized microsomes; BS, Blue Sepharose; GF, gel filtration; DE, DEAE ion exchange; HA, hydroxyapatite; HI, hydrophobic interaction chromatography; IF, isoelectric focusing; MQ, Mono Q (Pharmacia); NG, non-denaturing gel electrophoresis; RP, reverse-phase HPLC; SE, size exclusion chromatography.

zymes although the synthase has been purified from isolated starch grains. *Table 4* shows example protocols successful for these enzymes. Sucrose phosphate synthase is also an important plant enzyme and an example of its purification is given (26).

5.2.2 Cell wall polysaccharides

There is a prevailing notion that the enzymes responsible for the synthesis of cell wall polysaccharides are notoriously difficult to purify. This probably stems from the difficulties encountered in the attempts to purify cellulose synthase over many years. Part of the problem is due to the fact that these are membrane-bound enzymes and any solubilization process immediately changes the environment that they act in. Also studies have shown that solubilization releases hydrolases that break down the sugar nucleotides thus effectively reducing the substrate available to the solubilized polysaccharide synthase leading to apparent loss of activity during this stage (27). However chromatic separation of these hydrolases from the UDP-sugar utilizing polysaccharide synthases improves recovery.

These enzymes are also examples where catalytic activity is affected by the relative aqueousness of the environment. The behaviour of four membrane-bound glycosyl transferases involved in cell wall polysaccharide synthesis has been studied in relation to the action of graded series of organic solvents on their activity and type of product formed (Bolwell, G. P., unpublished data). Relative inhibition was in direct relationship to the hydrophilicity of the product where observed for some solvents. This was in the order of arabinan synthase > callose synthase > xylan synthase > 1–4, beta-glucan synthase. The first two were always inhibited while the latter two showed significant increases in incorporation in the presence of other solvents. A graded series of primary alcohols were much more effective in enhancing activity than acetone, ethyl acetate, and dimethyl formamide. In the presence of the most effective solvent, ethanol, the glucan synthase showed a fourfold higher 1–4, beta-glucan synthase activity with the virtual suppression of callose synthase activity. The products showed similar size and properties to those synthesized by the membrane-bound enzyme.

HPLC has been used to purify these solubilized glycosyl transferases. The key point to note is that the volume of the solubilized microsomes must be kept low and thus the protein concentration at maximum. It is an advantage to use micro-ultracentrifugation to deal with these small volumes and to produce a soluble preparation clean enough to load onto HPLC columns. It has proven possible to purify three glycosyl transferases from French bean but from tissues in different developmental states, by a two step protocol of DEAE ion exchange and size exclusion chromatography (*Table 4*). Other workers have used gel-based methods to identify protein bands corresponding to the glycosyl transferase. These have involved autoradiography of gels containing proteins tagged with [14C] azido-labelled compounds under denaturing conditions (18) or following radioactive product synthesis by proteins separated on native gels (17).

6 Enzymes of secondary metabolism

Plants are characterized by their extensive secondary metabolism. This is a subject of great commercial interest and is a target for genetic manipulation. While genes have been isolated by mutagenesis and complementation, there is still a place for cloning via the protein purification route.

6.1 Phenolic metabolism

Great strides forward have been made in the understanding of phenolic biosynthesis in plants. Plants utilize phenylpropanoids such as ferulic acid to strengthen primary cell walls by the formations of cross-linkages and secondary cell walls have extensive lignin polymers as part of their structure. Most of the enzymes contributing to these pathways have been purified (29, 30). Likewise the pathway to the flavonoids is completely known (34, 35). However much of the approach to the identification of genes coding for the enzymes of many of the individual steps has come from a molecular genetics approach. This has been possible because some of the enzymes belong to families, i.e. cytochrome P450 and oxoglutarate dependent dioxygenases. Functionality for the enzyme coded for by the cloned DNA has subsequently been established by transgenic methods. All these pathways contain enzymes that are soluble or membrane-bound.

6.1.1 Phenylpropanoids and lignin

Tissue cultures have been especially useful in characterizing the enzymes of phenylpropanoid metabolism. The phenylpropanoids and the B-ring of flavonoids are derived from phenylalanine and the shikimic acid pathway. The shikimate pathway is chloroplastic, similar to that in bacteria and is known in its entirety, with all the genes of the pathway now cloned. Phenylalanine is deaminated to cinnamic acid, the first intermediate of the phenylpropanoid pathway. The enzyme responsible, phenylalanine ammonia-lyase (PAL), is one of the best characterized enzymes of secondary metabolism in plants and has been purified from many plant sources. The pathway from the product of the PAL reaction, cinnamic acid, leading to the formation of the B-ring of the flavonoids shares part of the phenylpropanoid pathway to other phenolics particularly lignin. All the proteins, with the exception of *p*-coumarate-3-hydroxylase and the cytochrome P450, ferulate hydroxylase, have been purified to homogeneity for the phenylpropanoid pathway together with the enzymes of the lignin branch pathway, cinnamoyl-CoA reductase and cinnamoyl alcohol dehydrogenase and the probable enzymes of polymerization, lignin peroxidase and laccase. Protocols are shown for these enzymes (*Tables 2* and *5*).

6.1.2 Flavonoids

Although the formation and interconversion of flavonoids is complex the enzymology is less so because of multiple specificity (34, 35). The synthesis of the flavonoids proceeds from *p*-coumaroyl-CoA and malonyl-CoA and has been

Table 5 Successful protocols for the purification of the enzymes of phenylpropanoid and lignin biosynthesis

Enzyme	Assay	1	2	3	4	5
Phenylalanine ammonia-lyase (31)	S	AS	GF	DE	CF	CF
Cinnamic acid 4-hydroxylase (20)	B	SM	DE	HA		
O-methyl transferase (29)	R	AS	DE	SA	HI	
HCA-CoA O-methyl transferase (29)	R	AS	QS	BS	HA	IF
CoA ligase (29)	S	AS	BS	RA	MQ	
Cinnamoyl-CoA reductase (32)	S	AS	RA	MQ	MR	
Cinnamyl alcohol dehydrogenase (33)	S	AS	BS	MQ	AD	

Key: B, binding spectra assay; S, spectroscopic assay; R, radiometric assay; AS, ammonium sulfate precipitation; SM, solubilized microsomes; AD, ADP–Sepharose; BS, Blue Sepharose; CF, chromatofocusing; DE, DEAE ion exchange; GF, gel filtration; HA, hydroxyapatite; HI, hydrophobic interaction chromatography; IF, isoelectric focusing; MQ, Mono Q (Pharmacia); MR, Mimetic Red (Affinity Chromatography Ltd.); QS, Q-Sepharose; RA, Red Agarose.

extensively reviewed with respect to biochemistry, molecular biology, and genetic aspects. The first enzyme of the flavonoid pathway, chalcone synthase, is also extremely well understood. It is one of a family of polyketide synthases found in plants and catalyses the condensation of three malonyl-CoA derived from lipid biosynthesis with one *p*-coumaroyl-CoA. It also has been purified from many plants and a typical purification scheme is shown (*Table 6*). The initial product, tetrahydroxychalcone does not normally accumulate and is converted by chalcone isomerase to the generic flavone naringenin. This compound is then hydroxylated by the dioxygenase flavanone 3-hydroxylase to give dihydro-kaempferol which occupies a central position in the pathways that branch from it and are the basis of species-specific accumulation of flavonols and antho-cyanins. Although the various combinations of genetically determined enzymes operate to manifest the intra- and inter-specific variations in colour, there are

Table 6 Purification of enzymes of flavonoid biosynthesis (34)

Enzyme	Assay	1	2	3	4	5	6
		PG		IE			
Chalcone isomerase	S	AS	IE	DL			
Flavanone 3-hydroxylase	R	AS	IE	GF	HA	MP	HI
Flavone synthase	R	AS	QS	GF	MP	HI	
Dihydroflavone 4-reductase	R	AS	QS	DL	MQ		
Isoflavone oxidoreductase	R	AS	QS	DL	GF		
Pterocarpan synthase	TC	AS	QS	DL	GF		
Glycosyl transferases	R	AS	GD	AC	MP	GF	

Key: S, spectroscopic assay; R, radiometric assay; TC, TLC assay; AS, ammonium sulfate precipitation; SM, solubilized microsomes; AC, affinity chromatography; AD, ADP–Sepharose; CF, chromatofocusing; DE, DEAE ion exchange; DL, dye ligand centrifugation; GF, gel filtration; HA, hydroxyapatite; HI, hydrophobic interaction chromatography; MP, MonoP (Pharmacia); MQ, Mono Q (Pharmacia); QS, Q-Sepharose; RA, Red Agarose.

actually a limited number of enzymes involved which show multiple specificity. Hydroxylation of the B-ring is catalysed by two cytochrome P450s which have not been purified directly. Other enzymes such as flavonol synthase are oxo-glutarate dependant dioxygenases. Purification schemes are shown for a number of enzymes involved in flavonoid biosynthesis (*Table 6*).

6.2 Terpenoids

6.2.1 Monoterpenes

Great strides forward have been made in the purification of enzymes involved in the commercially important natural products which include menthol, limonene, and carvone. These are derived from geranyl pyrophosphate by cyclization and ring modification. 4S-limonene synthase has been purified from Mentha species (*Table 7*). Two cytochrome P450s are responsible for regio-specific hydroxylation of the generic compound limonene and have not been directly purified (8).

6.2.2 Carotenoids

Carotenoids are another example where the gene cloning has proceeded faster than the characterization of the enzymes. This has been aided by the availability of fungal mutants which allowed the cloning of the homologous plant gene. However, there are examples of purification of carotenogenic enzymes from plant sources. Although there are a number of intermediates identified in carotenoid biosynthesis, the enzymology is less complex since some of them catalyse multiple steps. Thus isopentenyl pyrophosphate isomerase, geranyl-geranyl pyrophosphate, and phytoene synthase have been purified from chloro-plast stroma or chromoplasts (*Table 7*) from Capsicum (36). Further enzymes of the carotenoid pathway, phytoene desaturase and xanthophyll synthase, have also been purified from chromoplast membranes from Capsicum (37, 38).

6.3 Alkaloids

Fewer enzymes of alkaloid biosynthesis have been purified to homogeneity (39, 40). The main sources have been tissue cultures. These developments have been

Table 7 Purification of enzymes of terpenoid and carotenoid biosynthesis

Enzyme	Assay	1	2	3	4	5	6
Isopentenyl isomerase (36)	R	CS	HI	HA	QS	GF	RA
Limonene synthase (8)	R	MA	MQ	HI			
Geranyl-geranyl synthase (36)	R	CS	HI	HI	HA	QS	
Phytoene synthase (36)	R	CS	HI	RA	QS		
Phytoene desaturase (37)	R	SM	QS	HA	MQ	MP	GF
Xanthophyll synthase (38)	R	SM	QS	AG	MP	GF	

Key: R, radiometric assay; CS, chloroplast stroma; SM, solubilized chloroplast membranes; AG, Affigel 501 (Bio-Rad); GF, gel filtration; HA, hydroxyapatite; HI, hydrophobic interaction chromatography; MA, Matrex Red (Amicon); MP, Mono P (Pharmacia); MQ, Mono Q (Pharmacia); QS, Q-Sepharose; RA, Red Agarose.

Table 8 Purification of enzymes involved in alkaloid biosynthesis (39, 40)

Enzyme	Assay	1	2	3	4	5
Strictosidine synthase	R	AS	GF	MA	CF	NG
Vindoline acetyltransferase	S	AS	GF	QS	HA	AC
Tryptophan decarboxylase	R	AS	DE	HA	GF	MQ
Tropinone reductase	S	AS	DE	HA	HI	MP
Hyoscyamine 6,beta-hydroxylase	R	AS	HI	DE	HI	HA

Key: S, spectroscopic assay; R, radiometric assay; AS, ammonium sulfate precipitation; solubilized micro-somes; AC, affinity chromatography (coenzyme A); CF, chromatofocusing; DE, DEAE ion exchange; GF, gel filtration; HA, hydroxyapatite; HI, hydrophobic interaction chromatography; MA, Matrex Red; MQ, Mono Q (Pharmacia); NG, non-denaturing gel electrophoresis; QS, Q-Sepharose.

possible due to screening of tissue culture sources for enrichment of the target enzymes. Advancement has also required the synthesis of radiolabelled sub-strates which are not commercially available. The range of alkaloids is vast but there are a few core pathways which lead to the generic intermediates which are modified in a species-specific manner. Most of the successes have come in understanding the biosynthesis of these key intermediates. Other reactions in-volve cytochrome P450s and are thus difficult enzymatically. Purification schemes are shown for key enzymes of indole and tropane alkaloid biosynthesis (*Table 8*). One cytochrome P450, berbamunine synthase (CYP80) has also been purified (23) (*Table 2*).

7 Preparation for Edman protein sequencing

Since the goal for many of the protein purifications has been to acquire protein sequence to underpin gene cloning, then some comments are probably worth making. This proteome analysis is meant to complement the immense genomic and EST analysis which is advanced in Arabidopsis, rice, and maize where a vast data base of primary sequences is accumulating. Proteome analysis allows identification of those DNA sequences that are expressed under particular developmental and environmental conditions for example. This matching is dependent on achieving meaningful amounts of sequence of good quality. Large scale blot technology is being applied for systematic sequencing programmes to produce overviews of the types of proteins being expressed in populations of cells, but the sequencing of purified proteins can be an add on bonus since it will identify particular expressed isoforms. Plant proteins do pose some problems in the final concentration steps before attempted sequencing (see *Protocol 7*).

Sample preparation should aim to remove unwanted peptides, minimize contamination of sample, and avoid any chemical modification. A trend towards the sequencing of fmol means sample preparation will have to be very precise. Accurate results can best be achieved on samples containing approximately 10–100 pmol. Samples smaller than this are difficult to sequence accurately. Sample quality must be established prior to sequencing for satisfactory results to be obtained. Thus protein or peptide samples must be substantially free of salts and detergents for good results. This is usually achieved by desalting into volatile solvent/buffer systems. Buffer salts in particular may give rise to problems by keeping the pH too low for complete reaction of PITC with primary and secondary amines in the sample. The purity of all reagents and the cleanliness of surfaces in contact with the sample are essential. Heavy metal contamination leads to low recoveries of some PTC amino acids, especially lysine.

The final step should aim to introduce volatile solvents. This is most conveniently carried out using sample concentrators or reverse-phase HPLC. Completely drying down the sample should be avoided and reduction to a small liquid volume often prevents solubility problems. The sample should ideally be in 30–90 µl aliquots which are then dried on to the specially prepared filter paper with nitrogen. If the sample is rather dilute, too many applications will cause uneven distribution of the sample with a tendency for the peptide to move to the edge of the filter paper. This will result in poor sequencing. Suitable solvents are, in order of preference, deionized water, 0.1% TFA, acetonitrile. The best choice is the solvent that the sample was previously dissolved in, if that is suitable. Liquid samples should be collected into Eppendorf tubes and stored at -20 to $-80\,^{\circ}C$. Samples stored under these conditions suffer less oxidation and sample loss is minimized.

Proteins can also be blotted onto PVDF paper (ABI Problott, Immobilon, etc.) and stained. The blotting buffer must be CAPS pH 11 and the proteins can be visualized with a variety of stains including Coomassie blue, amido black, Ponceau S, etc. They can be excised and sequenced using a blot cartridge.

Tris, pyridine, glycine, bicine, amino-sugars, polybuffers, ampholytes and some detergents, phospholipids, carbohydrates, and nucleic acids should be removed or avoided. Solutions of urea, guanidium–HCl, primary and secondary amines should also be avoided. Phosphate buffers and buffers containing ammonium salts, particularly the bicarbonate are also detrimental to optimizing sequence yields. Non-ionic detergents, such as Triton, in excess of 1%, if loaded with the sample will cause foaming in the transfer lines, and affect delivery of the reagents. Ionic detergents, such as SDS will cause problems in the same way, if in excess of 1%.

References

1. Lea, P. J. (1990). *Methods in plant biochemistry*. Vol. 3. *Enzymes of primary metabolism*. Academic Press, London.
2. Lea, P. J. (1993). *Methods in plant biochemistry*. Vol. 9. *Enzymes of secondary metabolism*. Academic Press, London.
3. Meyer, K., Cusmano, J. C., Somerville, C. R., and Chapple, C. C. S. (1996). *Proc. Natl. Acad. Sci. USA*, **93**, 6869.
4. Pear, J. R., Kawagoe, Y., Schreckengost, W. E., Delmer, D. P., and Stalker, D. M. (1996). *Proc. Natl. Acad. Sci. USA*, **93**, 12637.
5. He, X. Z. and Dixon, R. A. (1996). *Arch. Biochem. Biophys.*, **336**, 121.

6. Davin, L. B., Wang, H.-B., Crowell, A. L., Bedgar, D. L., Martin, D. M., Sarkanen, S., *et al.* (1997). *Science*, **275**, 362.

7. Kowalski, S. P., Eanetta, N. T., Hirzel, A. T., and Steffens, J. C. (1992). *Plant Physiol.*, **100**, 677.

8. Alonso, W. R., Rajaonarivony, J. I. M., Gershenzon, J., and Croteau, R. (1992). *J. Biol. Chem.*, **267**, 7582.

9. Gershenzon, J., McCaskill, D., Rajaonarivony, J. I. M., Mihaliak, C., Karp, F., and Croteau, R. (1992). *Anal. Biochem.*, **200**, 130.

10. Partington, J. C. and Bolwell, G. P. (1996). *Phytochemistry*, **42**, 1499.

11. Robertson, D., Mitchell, G. P., Gilroy, J. S., Gerrish, C., Bolwell, G. P., and Slabas, A. R. (1997). *J. Biol. Chem.*, **272**, 15841.

12. Zimmerlin, A., Wojtaszek, P., and Bolwell, G. P. (1994). *Biochem. J.*, **299**, 747.

13. Driouich, A., Laine, A.-C., Vian, B., and Faye, L. (1992). *Plant J.*, **2**, 13.

14. Wojtaszek, P. and Bolwell G. P. (1995). *Plant Physiol.*, **108**, 1001.

15. Bao, W., Whetten, R., O'Malley, D., and Sederoff, R. (1993). *Science*, **260**, 672.

16. Robertson, D., Smith, C., and Bolwell, G. P. (1996). *Biochem. J.*, **313**, 311.

17. Hanna, R., Brummell, D. A., Camirand, A., Hensel, A., Russell, E. F., and Maclachlan, G. A. (1991). *Arch. Biochem. Biophys.*, **290**, 7.

18. Delmer, D. P., Solomon, M., and Read, S. M. (1991). *Plant Physiol.*, **95**, 556.

19. O'Keefe, D. P. and Leto, K. J. (1989). *Plant Physiol.*, **89**, 1141.

20. Gabriac, B., Werck-Reichhart, D., Teutsch, H., and Durst, F. (1991). *Arch. Biochem. Biophys.*, **288**, 302.

21. Song, W.-C. and Brash, A. R. (1991). *Science*, **253**, 781.

22. Sibbesen, O., Koch, B., Halkier, B., and Moller, B. L. (1994). *Proc. Natl. Acad. Sci. USA*, **91**, 9740.

23. Stadler, R. and Zenk, M. H. (1993). *J. Biol. Chem.*, **268**, 823.

24. Leagood, R. C. (1990). In *Methods in plant biochemistry*, Vol. 3 (ed. P. J. Lea), p. 15. Academic Press, London.

25. Smith, A. M. (1990). In *Methods in plant biochemistry*, Vol. 3 (ed. P. J. Lea), p. 93. Academic Press, London.

26. Bruneau, J.-M., Worrell, A. C., Cambou, B., Lando, D., and Voelker, T. A. (1991). *Plant Physiol.*, **96**, 473.

27. Rodgers, M. W. and Bolwell, G. P. (1992). *Biochem. J.*, **288**, 817.

28. McCormack, B., Gregory, A. C. E., Kerry, M. E., Smith, C., and Bolwell, G. P. (1997). *Planta*, **203**, 196.

29. Strack, D. and Mock, H.-P. (1993). In *Methods in plant biochemistry*, Vol. 9 (ed. P. J. Lea), p. 45. Academic Press, London.

30. Whetten, R. and Sederoff, R. R. (1995). *Plant Cell*, **7**, 1001.

31. Bolwell, G. P., Bell, J. N., Cramer, C. L., Schuch, W., Lamb, C. J., and Dixon, R. A. (1985). *Eur. J. Biochem.*, **149**, 411.

32. Goffner, D., Campbell, M. M., Campargue, C., Clastre, M., Borderies, G., Boudet, A., *et al.* (1994). *Plant Physiol.*, **106**, 625.

33. Halpin, C., Knight, M. E., Grima-Pettenai, J., Goffner, D., Boudet, A., and Schuch, W. (1992). *Plant Physiol.*, **98**, 12.

34. Ibrahim, R. K. and Varin, L. (1993). In *Methods in plant biochemistry*, Vol. 9 (ed. P. J. Lea), p. 99. Academic Press, London.

35. Holton, T. A. and Cornish, E. C. (1995). *Plant Cell*, **7**, 1071.

36. Camara, B. (1993). In *Methods in enzymology*, Vol. 214 (ed. L. Packer) Academic Press, San Diego. p. 274.

37. Hugueny, P., Romer, S., Kuntz, M., and Camara, B. (1992). *Eur. J. Biochem.*, **209**, 399.

38. Bouvier, F., Hugueny, P., d'Harlingue, A., Kuntz, M., and Camara, B. (1994). *Plant J.*, **6**, 45.

39. De Luca, V. (1993). In *Methods in plant biochemistry*, Vol. 9 (ed. P. J. Lea), p. 345. Academic Press, London.
40. Hashimoto, T. and Yamada, Y. (1993). In *Methods in plant biochemistry*, Vol. 9 (ed. P. J. Lea), p. 369. Academic Press, London.

List of suppliers

Actigen Ltd., 5 Signet Court, Suranns Road, Cambridge CB5 8LA, UK.

Affinity Chromatography Ltd., 307 Huntington Road, Girton, Cambridge CB3 0JX, UK.

Amersham Pharmacia Biotech UK Ltd., Amersham Place, Little Chalfont, Buckinghamshire HP7 9NA
Tel: 0870 606 1921
Fax: 01494 544350
URL: http://www.apbiotech.com

Anderman and Co. Ltd., 145 London Road, Kingston-upon-Thames, Surrey KT2 6NH, UK.
Tel: 0181 541 0035
Fax: 0181 541 0623

Applied Biosystems,
Asaty Chemical Co. Ltd., The Imperial Tower 18F 1–1–1 Uchi Salwaicho, Chiyoda-ku, Tokyo 100, Japan.

BDH, see Merck Ltd.

Beckman Coulter (UK) Ltd., Oakley Court, Kingsmead Business Park, London Road, High Wycombe, Buckinghamshire HP11 1JU, UK.
Tel: 01494 441181
Fax: 01494 447558
URL: http://www.beckman.com

Beckman Coulter Inc., 4300 N Harbor Boulevard, PO Box 3100, Fullerton, CA 92834-3100, USA.
Tel: 001 714 871 4848
Fax: 001 714 773 8283
URL: http://www.beckman.com

Becton Dickinson and Co., 21 Between Towns Road, Cowley, Oxford OX4 3LY, UK.
Tel: 01865 748844   Fax: 01865 781627
URL: http://www.bd.com
Becton Dickinson and Co., 1 Becton Drive, Franklin Lakes, NJ 07417-1883, USA.
Tel: 001 201 847 6800
URL: http://www.bd.com

Bio 101 Inc., c/o Anachem Ltd., Anachem House, 20 Charles Street, Luton, Bedfordshire LU2 0EB, UK.
Tel: 01582 456666   Fax: 01582 391768
URL: http://www.anachem.co.uk
Bio 101 Inc., PO Box 2284, La Jolla, CA 92038-2284, USA.
Tel: 001 760 598 7299
Fax: 001 760 598 0116
URL: http://www.bio101.com

Bio-Rad Laboratories Ltd., Bio-Rad House, Maylands Avenue, Hemel Hempstead, Hertfordshire HP2 7TD, UK.
Tel: 0181 328 2000
Fax: 0181 328 2550
URL: http://www.bio-rad.com

Bio-Rad Laboratories Ltd., Division Headquarters, 1000 Alfred Noble Drive, Hercules, CA 94547, USA.
Tel: 001 510 724 7000
Fax: 001 510 741 5817
URL: http://www.bio-rad.com

Biospec Products, Bartlesville, OK, USA.

Bioprocessing Ltd., Consett, Durham DH8 6TJ, UK.

CP Instrument Co. Ltd., PO Box 22, Bishop Stortford, Hertfordshire CM23 3DX, UK.
Tel: 01279 757711
Fax: 01279 755785
URL: http://www.cpinstrument.co.uk

Dupont (UK) Ltd., Industrial Products Division, Wedgwood Way, Stevenage, Hertfordshire SG1 4QN, UK.
Tel: 01438 734000
Fax: 01438 734382
URL: http://www.dupont.com
Dupont Co. (Biotechnology Systems Division), PO Box 80024, Wilmington, DE 19880-002, USA.
Tel: 001 302 774 1000
Fax: 001 302 774 7321
URL: http://www.dupont.com

Dyax Corporation, Cambridge, UK.

Eastman Chemical Co., 100 North Eastman Road, PO Box 511, Kingsport, TN 37662-5075, USA.
Tel: 001 423 229 2000
URL: http://www.eastman.com

Fisher Scientific UK Ltd., Bishop Meadow Road, Loughborough, Leicestershire LE11 5RG, UK.
Tel: 01509 231166
Fax: 01509 231893
URL: http://www.fisher.co.uk

Fisher Scientific, Fisher Research, 2761 Walnut Avenue, Tustin, CA 92780, USA.
Tel: 001 714 669 4600
Fax: 001 714 669 1613
URL: http://www.fishersci.com
Fisher Scientific, Pittsburgh, PA, USA.

Fluka, PO Box 2060, Milwaukee, WI 53201, USA.
Tel: 001 414 273 5013
Fax: 001 414 2734979
URL: http://www.sigma-aldrich.com
Fluka Chemical Co. Ltd., PO Box 260, CH-9471, Buchs, Switzerland.
Tel: 0041 81 745 2828
Fax: 0041 81 756 5449
URL: http://www.sigma-aldrich.com

FMC Corp., Philadelphia, PA, USA.

Hampton Research, Customer Service Department 25431 Capot Road, Suite 205, Laguna Hills, CA 92653ñ5527, USA.
Tel: (741) 699ñ1040
URL: http://www.hamptonresearch.com

Hybaid Ltd., Action Court, Ashford Road, Ashford, Middlesex TW15 1XB, UK.
Tel: 01784 425000
Fax: 01784 248085
URL: http://www.hybaid.com
Hybaid US, 8 East Forge Parkway, Franklin, MA 02038, USA.
Tel: 001 508 541 6918
Fax: 001 508 541 3041
URL: http://www.hybaid.com

HyClone Laboratories, 1725 South HyClone Road, Logan, UT 84321, USA.
Tel: 001 435 753 4584
Fax: 001 435 753 4589
URL: http://www.hyclone.com

Invitrogen Corp., 1600 Faraday Avenue, Carlsbad, CA 92008, USA.
Tel: 001 760 603 7200
Fax: 001 760 603 7201
URL: http://www.invitrogen.com
Invitrogen BV, PO Box 2312, 9704 CH Groningen, The Netherlands.
Tel: 00800 5345 5345
Fax: 00800 7890 7890
URL: http://www.invitrogen.com

Life Technologies Ltd., PO Box 35, Free Fountain Drive, Incsinnan Business Park, Paisley PA4 9RF, UK.
Tel: 0800 269210
Fax: 0800 838380
URL: http://www.lifetech.com
Life Technologies Inc., 9800 Medical Center Drive, Rockville, MD 20850, USA.
Tel: 001 301 610 8000
URL: http://www.lifetech.com

Merck Ltd., Merck House, Poole, Dorset BH15 1TD, UK.
Tel: 01202 669700
Fax: 01202 666536

Merck Sharp & Dohme, Research Laboratories, Neuroscience Research Centre, Terlings Park, Harlow, Essex CM20 2QR, UK.
URL: http://www.msd-nrc.co.uk
MSD Sharp and Dohme GmbH, Lindenplatz 1, D-85540, Haar, Germany.
URL: http://www.msd-deutschland.com

Millipore (UK) Ltd., The Boulevard, Blackmoor Lane, Watford, Hertfordshire WD1 8YW, UK.
Tel: 01923 816375
Fax: 01923 818297
URL: http://www.millipore.com/local/UK.htm
Millipore Corp., 80 Ashby Road, Bedford, MA 01730, USA.
Tel: 001 800 645 5476
Fax: 001 800 645 5439
URL: http://www.millipore.com

Molecular Devices Inc., 135 Wharfedale Road, Winnash, Wokingham RG41 5RB, UK.

New England Biolabs, 32 Tozer Road, Beverley, MA 01915-5510, USA.
Tel: 001 978 927 5054

Nikon Inc., 1300 Walt Whitman Road, Melville, NY 11747-3064, USA.
Tel: 001 516 547 4200
Fax: 001 516 547 0299
URL: http://www.nikonusa.com
Nikon Corp., Fuji Building, 2-3, 3-chome, Marunouchi, Chiyoda-ku, Tokyo 100, Japan.
Tel: 00813 3214 5311
Fax: 00813 3201 5856
URL: http://www.nikon.co.jp/main/index_e.htm

Nycomed Amersham plc, Amersham Place, Little Chalfont, Buckinghamshire HP7 9NA, UK.
Tel: 01494 544000
Fax: 01494 542266
URL: http://www.amersham.co.uk
Nycomed Amersham, 101 Carnegie Center, Princeton, NJ 08540, USA.
Tel: 001 609 514 6000
URL: http://www.amersham.co.uk

Perkin Elmer Ltd., Post Office Lane, Beaconsfield, Buckinghamshire HP9 1QA, UK.
Tel: 01494 676161
URL: http://www.perkin-elmer.com

Pharmacia and Upjohn Ltd., Davy Avenue, Knowlhill, Milton Keynes, Buckinghamshire MK5 8PH, UK.
Tel: 01908 661101
Fax: 01908 690091
URL: http://www.eu.pnu.com
Pharmacia Biotech, Uppasala, Sweden.

Pierce Chemicals, Rockford, IL, USA.

Polysciences, Warrington, PA, USA.

Promega UK Ltd., Delta House, Chilworth Research Centre, Southampton SO16 7NS, UK.
Tel: 0800 378994
Fax: 0800 181037
URL: http://www.promega.com
Promega Corp., 2800 Woods Hollow Road, Madison, WI 53711-5399, USA.
Tel: 001 608 274 4330
Fax: 001 608 277 2516
URL: http://www.promega.com

Qiagen UK Ltd., Boundary Court, Gatwick Road, Crawley, West Sussex RH10 2AX, UK.
Tel: 01293 422911
Fax: 01293 422922
URL: http://www.qiagen.com
Qiagen Inc., 28159 Avenue Stanford, Valencia, CA 91355, USA.
Tel: 001 800 426 8157
Fax: 001 800 718 2056
URL: http://www.qiagen.com

Roche Diagnostics Ltd., Bell Lane, Lewes, East Sussex BN7 1LG, UK.
Tel: 01273 484644   Fax: 01273 480266
URL: http://www.roche.com
Roche Diagnostics Corp., 9115 Hague Road, PO Box 50457, Indianapolis, IN 46256, USA.
Tel: 001 317 845 2358
Fax: 001 317 576 2126
URL: http://www.roche.com
Roche Diagnostics GmbH, Sandhoferstrasse 116, 68305 Mannheim, Germany.
Tel: 0049 621 759 4747
Fax: 0049 621 759 4002
URL: http://www.roche.com

Rohm and Haas, Philadelphia, PA, USA.

Schleicher and Schuell Inc., Keene, NH 03431A, USA.
Tel: 001 603 357 2398

Scigen Ltd., Sittingbourne, Kent ME9 8AQ, UK.

Shandon Scientific Ltd., 93-96 Chadwick Road, Astmoor, Runcorn, Cheshire WA7 1PR, UK.
Tel: 01928 566611
URL: http://www.shandon.com

Sigma-Aldrich Co. Ltd., The Old Brickyard, New Road, Gillingham, Dorset XP8 4XT, UK.
Tel: 01747 822211
Fax: 01747 823779
URL: http://www.sigma-aldrich.com
Sigma-Aldrich Co. Ltd., Fancy Road, Poole, Dorset BH12 4QH, UK.
Tel: 01202 722114
Fax: 01202 715460
URL: http://www.sigma-aldrich.com
Sigma Chemical Co., PO Box 14508, St Louis, MO 63178, USA.
Tel: 001 314 771 5765
Fax: 001 314 771 5757
URL: http://www.sigma-aldrich.com

Stratagene Inc., 11011 North Torrey Pines Road, La Jolla, CA 92037, USA.
Tel: 001 858 535 5400
URL: http://www.stratagene.com
Stratagene Europe, Gebouw California, Hogehilweg 15, 1101 CB Amsterdam Zuidoost, The Netherlands.
Tel: 00800 9100 9100
URL: http://www.stratagene.com

United States Biochemical, PO Box 22400, Cleveland, OH 44122, USA.
Tel: 001 216 464 9277

Vertrieb & Kundendienst, Häka, Buttermaschinen GmbH, Stutensee, Germany.

Whatman International Ltd., Springfield Mill, James Whatman Way, Maidstone, Kent ME14 21E UK.
Tel: +44(0)1622 692022
Fax: +44(0)1622 691425

Index